George Duncan, Robert Koch, William Tennant Gairdner

Professor Koch on the Bacteriological Diagnosis of Cholera and Water-filtration

George Duncan, Robert Koch, William Tennant Gairdner

Professor Koch on the Bacteriological Diagnosis of Cholera and Water-filtration

ISBN/EAN: 9783337258542

Printed in Europe, USA, Canada, Australia, Japan

Cover: Foto ©berggeist007 / pixelio.de

More available books at **www.hansebooks.com**

PROFESSOR KOCH

ON THE

BACTERIOLOGICAL DIAGNOSIS OF CHOLERA,

WATER-FILTRATION AND CHOLERA,

AND THE

CHOLERA IN GERMANY DURING THE WINTER
OF 1892-93.

TRANSLATED BY
GEORGE DUNCAN, M.A.

WITH

PREFATORY NOTE BY W. T. GAIRDNER, M.D., LL.D., F.R.S.,
PROFESSOR OF THE PRACTICE OF MEDICINE, GLASGOW UNIVERSITY;
PRESIDENT OF THE ROYAL COLLEGE OF PHYSICIANS,
EDINBURGH, ETC., ETC.

NEW YORK :
WILLIAM R. JENKINS,
851 AND 853 SIXTH AVENUE. (48th Street)

EDINBURGH: DAVID DOUGLAS.
1895.

FOREWORD.

THE chief, and perhaps the only, reason which has led to my being asked to prefix a few words to this book, may be the fact that on perusing the translation, as originally published in the "Scotsman" newspaper, I had expressed very strongly, both to the publisher and to Councillor Pollard, chairman of the Edinburgh Public Health Committee, the desire that the work of Professor Koch might appear in English in a form for more permanent reference.

There are many persons in this country, medical and other, who have little confidence in Professor Koch; and it may be admitted that the failure of his therapeutic results in the instance of the toxic inoculation for tubercular diseases has been a great trial to those who believe, as I do, that he is not only one of the most distinguished, but also one of the most trustworthy authorities of this century on all subjects connected with bacteriology in its applications to diagnosis, and also to the prevention of disease.

This work will, I believe, at once confirm Professor Koch's reputation in this respect with those who have hitherto believed in him; and will also, perhaps, win him some converts even from among those who have, (in my opinion quite wrongfully,) disparaged the whole of his work on account of the imperfections of a small part of it.

It ought to be interesting, at all events, to medical men in this country to find the admirable work begun by the late Dr Snow, so long ago as 1849, (during the second great epidemic of cholera,) carried on to an issue so precise and convincing as in the cases of Hamburg, Altona and Nietleben epidemics with the help of bacteriological diagnosis.

Since Dr Snow's researches were published and adopted by the registrar-general in England, there has never been much doubt among us as to the water-communication of the choleraic infection, the evidence of which seemed to go on accumulating as the incidence of the disease, in respect to particular places, was more and more studied, and the severity of local epidemics was found to be strictly in accordance with the presence of dangerous impurities in the water-supply.

In 1862, the argument had been so far advanced that I was able, in a volume entitled "Public Health in relation to Air and Water," to develop it in great detail in its relation to the old "contagion" and other theories of the origin and distribution of cholera epidemics.

Sir John Simon, in his numerous reports as medical officer of the Privy Council, has insisted on the like conclusions, and has expressively indicated cholera as one of the "excremental diseases," or "filth fevers." The arguments of Von Pettenkofer and of the Munich School have not, so far as I have observed, taken deep root among Sanitarians in this country; but, nevertheless, there may be some who may not at first be disposed to coincide with the highly controversial attitude of Dr Koch in rebutting the ideas of his opponents.

It cannot fail to be conceded, however, that in respect of luminous and accurate setting forth of his facts, and also of firm and logical statement of his arguments, the present treatise of Koch is simply a model of sanitary research, and, as such, I commend it to the careful perusal of all who have any special interest in the subject.

W. T. GAIRDNER.

UNIVERSITY, GLASGOW,
 17th October, 1894.

TRANSLATOR'S PREFACE.

IN laying this translation before the British public, I think it right to state that I received very valuable assistance in preparing it from Professor Koch and from Dr Frosch, Professor Pfeiffer, and Professor Pfuhl, who work under him in the Institute for Infectious Diseases here. I have also to express my acknowledgments to the Editor of the "Scotsman" for his liberality in permitting the reprinting of this translation, which first appeared in the columns of that paper. Last, but not least, I have to mention my debt to my public-spirited friend, Councillor Pollard, Chairman of the Health-Committee of Edinburgh, who succeeded in interesting first the Editor of the "Scotsman," and then the publisher of this volume, in the enterprise.

GEORGE DUNCAN.

BERLIN, 20*th October*, 1894.

CONTENTS.

	PAGE
ON THE PRESENT STATE OF THE BACTERIOLOGICAL DIAGNOSIS OF CHOLERA,	1
1. THE MICROSCOPIC INVESTIGATION,	6
2. THE PEPTONE-CULTIVATION,	8
3. THE GELATINE-PLATE-CULTIVATION,	11
4. THE AGAR-PLATE-CULTIVATION,	14
5. THE CHOLERA-RED REACTION (INDOL-REACTION),	15
6. EXPERIMENTS ON ANIMALS,	16
WATER-FILTRATION AND CHOLERA,	24
THE CHOLERA IN GERMANY DURING THE WINTER OF 1892–93,	64
1. THE AFTER-EPIDEMIC IN HAMBURG,	65
2. THE WINTER EPIDEMIC IN ALTONA,	85
3. THE CHOLERA-EPIDEMIC IN THE LUNATIC-ASYLUM AT NIETLE-BEN NEAR HALLE,	101
1. THE EPIDEMIC OF 1850,	114
2. THE EPIDEMIC OF 1866,	116
3. THE EPIDEMIC OF 1893,	119

On the Present State of the Bacteriological Diagnosis of Cholera.

Soon after the cholera-bacteria and their relations to cholera became known, opposition arose in various quarters to the opinion that these bacteria were the exclusive concomitants of cholera, and might be turned to account in its diagnosis. I may remind the reader that people professed to have found the same bacteria in the mucus of the teeth of healthy persons, in the water of districts free of cholera, in cholera nostras, etc.; but I may also remind him that these statements were very soon refuted. As moreover the regular occurrence of the cholera-bacteria in genuine Asiatic cholera was confirmed a thousand times over in several epidemics which developed in the following years in France, Italy, Spain, and South America, and as all the experience of the present epidemic has taught the same, we may now, I think, regard it as *an ascertained fact that the cholera-bacteria are the inseparable concomitants of Asiatic cholera, and that the demonstration of their presence is an infallible proof of the presence of this disease.** So far as I know, this proposition is no longer dis-

* This does not imply, however, that the absence of the cholera-bacteria, or rather the failure to find them, in a case suspected to be one of cholera, proves under all circumstances that the disease in question is not cholera. In cholera, just as in other infectious diseases caused by micro-organisms, cases may occur which one must, owing to their other symptoms, regard as indubitable cases of cholera, in which, however, owing to the defective qualification of the investigator, or to the circumstance that they are investigated at an unsuitable time, cholera-bacteria are not found.

puted by anyone whose opinion deserves serious consideration. That in my eyes, and in the eyes of everyone who possesses an adequate knowledge of bacteriology and of the nature of infectious diseases, the proof of the specific character of the cholera-bacteria is at the same time a proof that they are the cause of cholera, need hardly, I think, be expressly stated at this time of day; and I do state it here once more only because, strange to say, there still are medical men who doubt the causative relation between cholera-bacteria and cholera, though they cannot give even the semblance of a proof of any other relation between these two inseparably connected things.

Whatever one may think of the cholera-bacteria, however, everybody will have nowadays to admit that Asiatic cholera must be present in every case in which one finds them, and that therefore the demonstration of their presence in doubtful cases is of the utmost diagnostic importance. In certain cases, as is well known, the clinical symptoms of Asiatic cholera cannot be distinguished with certainty from those of cholera nostras, infantile cholera, certain forms of peritonitis, and poisoning by arsenic and some organic poisons. Only its high mortality and its infectious character, which reveals itself by the development of groups of cases and by spreading to other places, enable one to distinguish Asiatic cholera from the above-named diseases. At the height of the epidemic, when there can no longer be any doubt of the infectious character of the disease, the bacteriological investigation is therefore not needed to enable one to recognise all but the slightest cases of diarrhœa with vomiting as cases of Asiatic cholera; though here also bacteriological investigation, carried on in as many cases as possible, would guard against not a few mistakes, and is necessary in doubtful cases. In hospitals too bacteriological investigation will have to be carried on in future on a larger scale, because, if for no other reason, it has been ascertained that the excrements may contain cholera-bacteria a good while after the actual attack of the disease, and it is not permissible to dismiss patients till they are free of cholera-bacteria. The proper field of bacteriological work, however, is the beginning and the end of an epidemic, when all depends on the correct judging of each individual case, and the swiftest possible prevention of danger to the neighbourhood. In former times the beginning and end of a local epidemic could hardly ever be recognised with the necessary certainty; they were to a certain extent veiled, so that one could indeed trace the rude lines of the epidemic proper, but generally lost the thread towards the beginning and the end. Hence too it came that the first measures were taken too late, and, when the epidemic began to abate, the efforts to combat it were stopped much too early. A material change has taken place in these respects. In the highly complicated net which cholera forms in its ways and in its spread only isolated threads still remain hidden to us; all the rest, to the tiniest off-shoots, lies clear and distinct before our eyes. Now at last we can oppose the pestilence step by step, and combat it when it is small and weak, that is, when the prospect of success is greatest; and the course the present epidemic in Germany has hitherto taken has shown the high utility of this kind of cholera-prophylaxis, directed against individual cases, beyond a doubt.

In order, however, completely to utilise the advantages of bacteriological diagnosis, it is absolutely necessary that it be swift and sure. For the spread of cholera, both in the place of its outbreak and from there to other places, is generally so rapid that the delay of the measures for some days, nay even for one day, may do most terrible and irreparable damage. Moreover bacteriological technics must be able to diagnose even the slightest cases of Asiatic cholera, which show hardly perceptible indications of morbid symptoms, but which the presence of cholera-bacilli proves to be cases of genuine cholera. That there are mild, sometimes very mild, cases of cholera, had been conjectured before, though it had not been possible really to prove it; but that the gradations in the intensity of the disease can go so far as the experience of the last epidemics actually shows, is a truth first taught by bacteriology, and it is not necessary, I suppose, to indicate in detail what a dangerous part such mildest cases may play in the spreading of cholera, and how important their detection must be.*

The method originally described by me, of which, as is well known, cultivation on gelatine plates forms the basis, has, as very soon appeared in the course of last year's epidemic, not always adequately fulfilled the demands just stated. As regards certainty, indeed, it left nothing to be desired, for not a single instance of its failure in ordinary cases in skilful and practised hands has ever come to my knowledge. Only in the above-mentioned mildest cases, in which the excreta contain very few cholera-bacteria, cases which have become accessible to us only by the methods to be described below, does the gelatine-plate-method not suffice. But the point in which the method hitherto in use left most to be desired was rapidity of demonstration. If one proceeded in the ordinary way, and used plates on which the colonies must have developed so far that they presented a characteristic aspect, and were large enough to admit of the making of microscopic preparations from a single colony, the investigation generally lasted, according to the conditions of temperature, from a day and a half to two days. In many cases, however, when the task was the first diagnosis in a place, a still longer period, varying from two to five days, elapsed before it was possible to declare with certainty whether the disease was Asiatic cholera or not. Such delays, however, are not to be ascribed to the method, but to the inexperience of the investigators. I could give a good number of such instances, but will restrict myself to one case, which has become the best known of all, and also led probably to the most serious

* These mildest cholera-cases, in which cholera-bacteria were found in the solid excreta of apparently healthy persons, occurred only among groups of persons who were equally exposed to infection, and some of whom had the disease in a severe form. Among persons who were not exposed to infection at all nothing of this kind has ever been found. One must therefore regard these cases as genuine cases of cholera, and cannot use them as arguments against the specific character of the cholera-bacteria.

consequences. On the 16th of August 1892 the first cholera-patient was taken to the Hamburg Hospital, and died there on the 17th. The autopsy took place on the 18th, and the bacteriological investigation was begun by Dr Rumpel on the same day.* It yielded no result, however, till the 21st of August, and would probably never have yielded any, had not the plates got into the hands of E. Fraenkel, who was able to express a definite opinion on the following day. I myself have seen several cholera-colonies on a gelatine plate of the third dilution of this case preserved in a refrigerator, and was able to convince myself that in this case also the diagnosis presented no special difficulties. If the excreta of the patient had been bacteriologically examined by an expert on the 16th of August, when the clinical symptoms were so suspicious that he was taken to an isolation-barrack, and that strict precautions were taken, a sure diagnosis would have been obtained on the 18th at latest. The excreta of a suspicious case were examined by Dr Weisser in Altona on the 19th of August, and by the 21st pure cultivations of cholera-bacteria were obtained from them, which Dr Weisser demonstrated in Berlin on the following day.

In the hands of a practised bacteriologist the bacteriological investigation of a case of cholera by the older method required, as already mentioned, about two days. The shortening of this so precious time, if possible, could not but be of the greatest value. This conviction has forced itself upon many bacteriologists who have had to do with cholera-investigations, and has induced not a few to endeavour to improve the method in this respect. Happily these efforts have not proved unavailing, and we now have essential

* German Medical Weekly, 1893, No. 7, page 161. It is unintelligible to me why Dr Rumpel states in his essay, in spaced out print (the German equivalent for italics), that neither I nor Dr Weisser "had anything to do with the determination of the cholera-diagnosis in Hamburg," as neither I nor Dr Weisser have ever claimed to have had anything to do with it. All I had to do with it was that I expressed my surprise that the cholera-diagnosis had taken so much time in Hamburg, whereas it had been made so quickly in Altona.

improvements at our disposal, of which it is not always possible to say with certainty to whom we owe them. Many have helped in the matter, one contributing a little, another more. It is to be hoped too that further improvements will be introduced in course of time. If I have nevertheless undertaken to give a general view of the present state of the diagnosis of cholera before its development can be said to be really complete, the reason is that those improvements are either not known yet at all or are inadequately known, and are used only in bacteriological laboratories; whereas it is highly desirable that they should as soon as possible be made accessible to the widest circles, in order that they may be usefully applied in the future combating of cholera.

To avoid unnecessary prolixity, I shall omit all that has in course of time proved less advantageous or superfluous and been therefore abandoned; I shall also abstain from giving the complete historical development of the method and pointing out the merits of each contributor, restricting myself to describing the method which is practised at present in the Institute for Infectious Diseases, and which has stood the test of large experience. To this method belong the following parts.

1. The Microscopic Investigation.

After the experience of the last epidemic I must now attach a much greater value to the microscopic investigation than before.

This part of the investigation consists in making a coverglass preparation of the object to be examined (the contents of the intestines of a corpse, or the excreta of a patient) in the well known manner; a mucus-flake taken from the fluid being, if at all possible, selected for this purpose. For staining it is advisable to use a dilution of Ziehl's solution of fuchsine.

According to the severity of the case and the stage of the disease, the cholera-bacteria in such preparations are wholly or nearly in pure cultivation, or mixed with the usual bacteria of the intestines, among which the bacterium coli predominates in all gradations down to cases in which curved little staves are no longer visible through the microscope.

If the cultivation of the cholera-bacilli is pure, or mixed only with less numerous bacteria coli, the cholera-bacteria generally lie together in characteristically formed groups at the places at which the mucus in the preparation is drawn out in the form of a thread. They form little heaps in which the single bacilli all have the same direction, so that it looks as if a little swarm of them were moving behind one another, like fish in slowly flowing water. Even in my first investigations this singular stratification of the cholera-bacteria in the mucus-flakes had struck me, and I published an illustration of it on the occasion of the first conference for the discussion of the cholera-question (26th July 1884).* Since then I have regularly drawn attention to it in lectures and demonstrations. There are also photographic representations, taken from my preparations, in Gaffky's report of the work of the commission sent to Egypt and India in 1883 for the investigation of cholera (Berlin, 1887), Plate 15, illustrations 7 and 8, and in C. Fränkel's and R. Pfeiffer's Microphotographic Atlas (1891), illustrations 84 and 85. I have never seen groupings bearing even the most distant resemblance to those here described in cases which further investigation showed not to be cases of Asiatic cholera. They are found exclusively in cholera, and I therefore regard them as so characteristic that one may with certainty pronounce the disease under investigation to be Asiatic cholera on the evidence of them alone. I even go a step further. If the peculiar grouping of the cholera-bacteria be wanting in microscopic preparations of excreta, but only bacteria coli be found along with numerous scattered bacteria which have the appearance of cholera-bacteria, then also one may conclude with certainty that the disease in question is Asiatic cholera. Only when the mixture of bacteria becomes more complicated does the microscopic diagnosis begin to be uncertain.

But, if I maintain that the diagnosis may be based on the

* German Medical Weekly, 1884, Nos. 32 and 32A.

microscopic investigation alone, I presuppose that the investigator possesses great practice and experience, and also an eye for the differences of the forms of bacteria, a qualification which I have missed in not a few cases even in practised bacteriologists. Whoever therefore wishes to turn the microscopic examination of cholera-objects to reliable account in diagnosis must first acquire the necessary practice by means of numerous suitable preparations, made if possible by himself, and sharpen his eye for the recognition of the morphological character of the cholera-bacteria. With this proviso the microscopic diagnosis is of the utmost importance, for, according to the experience gathered in the Institute for Infectious Diseases during the last epidemic, it alone formed the basis of a sure diagnosis in almost 50 per cent. of the cases the excreta of which were sent to the Institute for investigation; that is, the Institute was thereby enabled to make the diagnosis and telegraph the result of the investigation to the sender in about half of all the cases only a few minutes after the arrival of the material. In all these cases of course the investigation was afterwards so far completed that no doubt of the provisional diagnosis remained, and it was not necessary in a single case to alter the opinion based on the microscopic investigation alone. Everyone who knows that cholera can never be dealt with too promptly can judge of the prophylactic importance of such rapid diagnosis. How useful it would be too if a great number of medical men were practised in microscopic diagnosis, I had occasion to observe not long ago in a case in which attempts at cultivation had been made for four days in vain, and in which the first microscopic preparations, as I was afterwards able to ascertain, would have enabled a skilled bacteriologist to make the correct diagnosis at once.

2. THE PEPTONE-CULTIVATION.

In the course of his investigation of the chemical reactions of cholera-cultivations * Dunham had found that the cholera-

* Periodical for Hygiene and Infectious Diseases, 1887, vol. ii., page 337.

bacteria multiplied very quickly at incubation-temperature in a sterilised solution of 1 per cent. of peptone, and 0·5 per cent. of common salt. After six hours the fluid had become perceptibly dimmer, and gave the well known cholera-red reaction with sulphuric acid. As cultivations in bouillon with and without peptone, treated in the same way, yielded much less favourable results, it followed that a pure solution of peptone, with a corresponding content of common salt, must be a specially good nutritive medium for the cholera-bacteria. Since then, however, this observation was made use of only when one wished to produce the cholera-red reaction in the purest possible form. It was not applied, so far as I know, in order to accelerate the demonstration of the cholera-bacteria in excreta, etc. During the last epidemic Dunbar of the Hygienic Institute in Hamburg used Dunham's discovery for the first time in investigating cholera-material, and developed it into a serviceable method. It is now applied in the following manner: One or more platinum loops of the excreta, or, if the latter contain mucus-flakes, several such flakes, are put into the sterilised 1 per cent. peptone-solution (which is contained in test-tubes), and then kept at a temperature of 37 degrees centigrade. The cholera-bacteria, as Hesse has so convincingly shown,* need oxygen in a very high degree; they strive towards the surface of the fluid, and multiply there, undisturbed by the other bacteria of the fæces, which, at first at least, remain more in the deeper layers of the fluid. If one takes droplets from the surface of the peptone-solution with the platinum loop as soon as the fluid shows the first traces of getting dim, and examines them under the microscope, and if there have been many cholera-bacteria in the seed-material, one often finds a pure cultivation of cholera-bacteria on the surface of the peptone-solution after the lapse of only six hours. If they have been less numerous in the material, they appear on the surface later, and more or less mixed with bacteria of the fæces (especially bacteria coli), so that the microscopic investigation may leave a doubt whether the

* In the Periodical for Hygiene and Infectious Diseases, vol. xiv.

curved bacteria found are cholera-bacteria or not. In all cases, however, the quantity of cholera-bacteria in the material under investigation, if it originally contained cholera-bacteria at all, is so largely increased by the peptone-cultivation that the further investigation and isolation, which would have been uncertain or without prospect of success with the gelatine-plate-cultivation and similar methods alone, are rendered much easier. For it has been observed, both in the Hygienic Institute in Hamburg and in the Institute for Infectious Diseases, that in a small number of mild cases, specially difficult owing to the very small quantity of cholera-bacteria present, in which nothing more could be found by the gelatine-plate-method, the peptone-cultivation still yielded positive results. My provisional explanation of this is, that in the plate-cultivation the bacteria of the fæces, predominant in such cases, overgrow the cholera-bacteria, which cannot evade them as in the peptone-fluid, and so greatly check their development that they remain stunted and invisible, not only to the naked eye but even when slightly magnified. At any rate the peptone-cultivation has shown itself so superior to the plate-method on this occasion that I should not like to be without it in any investigation. It justifies a definite opinion only when the cholera-bacteria are found in pure cultivation on the surface of the fluid. In other cases it offers only the advantage—an indispensable one indeed—of enrichment in cholera-bacilli. The best time to examine the peptone-solution is six to twelve hours after the putting in of the seed-material; sometimes one must wait longer. It is necessary to examine a specimen from time to time, in order to hit the moment of the maximum-development of the cholera-bacteria. Afterwards they are overgrown and supplanted by other bacteria, even in the upper strata of the fluid, and it may happen that, if the examination be delayed too long, they can no longer be found.

It must be added that, according to the experience obtained in the Institute for Infectious Diseases, it is advantageous to increase the addition of common salt to one per cent., and to make the fluid strongly alkaline, if it is not so already. It is also necessary to note that the peptone-preparations on sale are not all suitable for the purpose in question. We prefer the preparation sold by Witte of Rostock. As the peptone of commerce, even when made in the same factory, does not always possess the same alkaline content, the addition of soda cannot be definitely stated.* It is therefore necessary, before using a particular kind of peptone for this purpose, to test its qualification to serve as a specially good nutritive medium for cholera-bacteria, and to determine the amount of soda to be added in order to obtain the most luxuriant cultivations.

3. THE GELATINE-PLATE-CULTIVATION.

The appearance of the cholera-colonies in gelatine plates is so characteristic, especially if they are present in preponderant number or even in pure cultivation, that this method, though, as has been mentioned, it is surpassed in subtlety by the peptone-cultivation, cannot be dispensed with. Peptone-cultivation and gelatine-plate-cultivation must supplement one another. By the peptone-cultivation the object is so enriched in cholera-bacteria in a few hours that the gelatine plate also, which without this aid would have brought only isolated colonies, or in certain cases none at all, to development, is now oversown with characteristic colonies.

In the technics of the gelatine-plate-method no change has been made. Three dilutions are prepared and poured into double shallow glass dishes in the well known manner described in all text-books of bacteriology. The warmer these dilutions are kept, the more quickly the colonies appear; one must therefore, in order to accelerate the process as much as possible, expose the plates to a temperature at which the gelatine remains just solid enough to prevent the excessive liquefaction of the cholera-colonies and the loss of their char-

* The peptone used in the Institute contains 0·025 $Na_2 CO_3$ per gramme (determined by titration, litmus-paper being applied as indicator).

acteristic appearance. If well prepared ten-per-cent. gelatine is used, this temperature is 22 degrees centigrade. Under these circumstances the colonies assume their characteristic aspect in fifteen to twenty hours. As a slight increase of the above-mentioned temperature is followed by the softening of the gelatine and the liquefaction of the colonies, it is absolutely necessary to keep the plates in an incubator set at 22 degrees, which does not allow greater fluctuations than half a degree above and below that point. If the cultivations are kept at too high a temperature, or if the gelatine is bad and becomes too soft at 22 degrees, the cholera-colonies liquefy the gelatine to a considerable extent, and they assume an aspect greatly resembling that of Finkler's bacteria. Unpractised investigators can therefore, if they have gone unskilfully to work, easily fall into the mistake of supposing that they have Finkler's instead of cholera-bacteria before them, a mistake which actually happened repeatedly during the last epidemic.

I avail myself of this opportunity to put an end to an error which has maintained itself in bacteriology for a series of years, and passed without criticism from one text-book into another, the supposition, namely, that Finkler's bacteria are connected in some way with so-called cholera nostras, and can be confounded with the bacteria of Asiatic cholera. Finkler's bacteria were originally cultivated from the several days old putrid excreta of a person suffering from diarrhœa with vomiting. Their discoverer was unable to prove their regular and exclusive occurrence in cholera nostras, and I have therefore from the first denied the importance claimed for them. Since then I have very often had occasion to examine cases of cholera nostras, but have never found Finkler's bacilli. Other practised bacteriologists have made the same experience. Finkler himself told me some time ago, by word of mouth, that he too had never found them again. The most comprehensive investigations were made in this direction on the occasion of the last epidemic, when thousands of cases suspected of being cholera-cases, among which of course there were numerous cases of cholera nostras, were bacteriologically investigated, but Finkler's bacteria were not conclusively demonstrated in a single case. Where they were allegedly found in Peter's clinic in Paris for instance, no expert can be in doubt for a moment that their discovery was due to egregious blunders on the part of a beginner in bacteriology. After such experience Finkler's bacteria, which have hitherto brought nothing but confusion, and even caused a mistake which led to very serious consequences in a case that has come to my knowledge, should really disappear at last from the scene, and cease to serve as a bugbear to incipient cholera-bacteriologists.

Under ordinary circumstances, that is, when fresh cholera-bacteria develop in properly prepared gelatine, and at not too low a temperature, the appearance of the colonies is that which has been repeatedly described and represented in photographs.* Deviations in the composition of the gelatine, slow development at a low temperature, may produce a different appearance, the right judging of which requires practice. Older cultivations, long recultivated in a laboratory, also yield a growth deviating more or less from the typical one. In the case of fresh cholera-bacteria I have only once seen such an abnormal growth on gelatine, and the distinguishing circumstance in this case is that the colonies possess a very slight inclination to liquefaction, and consequently spread out at first in forms resembling plates or shields. In all other qualities, especially in the essential ones afterwards to be described, these bacteria agreed entirely with the ordinary cholera-bacteria, and must therefore be regarded as genuine cholera-bacteria. In consideration of the fact that such deviations in growth on gelatine plates have been observed, though, as I expressly remark, only quite exceptionally, it is necessary completely to clear up apparently doubtful cases by applying other criteria.

* Compare Riedel on the Cholera of 1887; Gaffky's Report of the Cholera-Expedition; Fraenkel's and Pfeiffer's Microphotographic Atlas.

4. THE AGAR-PLATE-CULTIVATION.

This method is, strictly speaking, only a modification of the preceding one, but differs from it in some essential points. The growth of the cholera-bacteria on agar is not so characteristic as that in gelatine, and one cannot designate them at once as cholera-colonies in virtue of their appearance alone. A practised eye, however, can distinguish cholera-colonies that have grown on agar from the ordinary bacteria of fæces and water with tolerable certainty. When they have developed on the surface of agar, they form moderately large colonies with a singular clear grey-brown transparent appearance, whereas almost all other bacteria here in question form less transparent colonies. In order to avoid mistakes, the investigator must examine the said colonies each time under the microscope, with a view to ascertaining whether they consist of bacteria which agree morphologically with the cholera-bacteria. If the appearance of these colonies to the naked eye were alone in question, the agar-plate-cultivation would offer no advantages whatever over the gelatine-plate-cultivation, and could not be turned to account in the further perfecting of bacteriological diagnosis. But the important gain we can derive from the agar-cultivation is due to the fact that it can be exposed to a high temperature (37 degrees centigrade), and therefore furnishes in eight to ten hours colonies of a size adapted to our further purposes. As it is necessary to this end to make the colonies grow on the surface and not in the interior of the agar layer, where they develop considerably more slowly and remain smaller, one does not mix the seed-material with the liquefied agar, as is always done in the gelatine-method, but pours the liquefied agar into double shallow glass dishes, lets it congeal, and then spreads the seed-material with a platinum loop on the surface of the agar. But as, further, freshly poured agar forms a thin layer of fluid on the surface, which would impede the development of isolated colonies, the double shallow glass dishes containing the agar must stand for some days in an incubator till the fluid has evaporated, before they can be used further. Of course the inoculated agar plates are placed in an incubator set at 37 to 38 degrees centigrade. Quite isolated cholera-colonies can hardly be singled out among many others on the agar plate, but, if they are only tolerably numerous, they strike the eye at once. For this reason the agar-cultivation is not suitable for seed-material which is very poor in cholera-bacteria; all the more so, however, after it has been enriched by the peptone-cultivation. We therefore use this method almost exclusively for the further completing of the peptone-cultivation. By means of the latter one obtains in six to ten hours a fluid rich in cholera-bacteria, out of which, after a further period of eight to ten hours, comparatively large colonies, that is, pure cultivations of cholera-bacteria, arise on agar. After being examined under the microscope, these can be used partly for the preparation of pure cultivations in peptone-solution, which in a short time render the cholera-red reaction possible, partly, if they are numerous and large enough, for experiments on animals.

5. THE CHOLERA-RED REACTION (INDOL-REACTION).

This reaction, which was discovered almost simultaneously by Bujwid and Dunham, takes place, as is well known, by the appearance of a red colouring on the addition of sulphuric acid in the cholera-cultivations, which contain indol and nitrous acid. Other bacteria also produce indol just as cholera-bacteria do; others, again, can reduce nitric acid to nitrous acid; perhaps there are also some which possess both these qualities in common with the cholera-bacteria; but none of the bacteria hitherto known which have a curved form, and might for this reason be confounded with the cholera-bacteria, yield indol and nitrous acid simultaneously in their cultivations, and these bacteria do not give the red reaction. For this reason we must attach a very high value to the cholera-red reaction as a means of distinguishing the cholera-bacteria from bacteria of similar form. But, if the reaction is to afford full certainty, the following precautions must be strictly observed. Above all one must get a suitable kind of peptone, for the different kinds do not all yield equally good results. The cause of this difference, as Bleisch has shown,* probably is that the kinds of peptone which do not give the reaction well are either too rich or too poor in nitrates. One must therefore either find a suitable peptone by preliminary experiments, and procure a sufficient quantity of it, or, as Bleisch proposes, add the necessary amount of nitrate to peptones poor in or free from nitrate, if one can get such. A second requisite for the reliability of the reaction is that the sulphuric acid used be absolutely free from nitrous acid. The third condition is that the reaction be effected only with a pure cultivation of cholera-bacteria. If this is not the case, and if the experiment succeeds, the objection that the indol or the nitrous acid emanates from other bacteria is not excluded.

As the reaction does not take place so equably and distinctly with cultivations which have grown in bouillon, even if it contains peptone, it should always be made with cultivations obtained in a peptone-solution.

6. Experiments on Animals.

It had been known for a good while that cultivations of cholera-bacteria could have a poisonous effect, if injected into the peritoneal cavity of guinea-pigs, but the statements made on the subject contradicted each other. It was first ascertained by the investigations of R. Pfeiffer † that the poison is contained mainly in the cholera-bacteria themselves, and that one obtains equable and reliable results, only if not fluid cultivations but cultivations grown on agar are injected in a definite quantity into the peritoneal cavity of guinea-pigs. Following Pfeiffer's example, one takes a loopful from the surface of the agar with a platinum loop, which can take about 1·5 milligramme of the cultivation, distributes it in one cubic

---

* Periodical for Hygiene and Infectious Diseases, vol. xiv.
† *Ibid.*, vols. xi. and xiv.

centimetre of sterilised bouillon, and injects it into the peritoneal cavity. For this little operation too one must of course train oneself by practice; for, when it is undertaken by unpractised hands, it very often happens that the fluid is injected not into the peritoneal cavity, but wholly or partly into the pierced intestine. The size of the animals experimented on is not indifferent; the larger they are, the larger the mortal dose must be. A platinum loop of the above-mentioned size may be regarded as a certainly mortal dose for a guinea-pig of 300 to 350 grammes. If the experiment is correctly gone about, it invariably succeeds, and the singular phenomena of poisoning, minutely described by Pfeiffer, soon appear, especially the rapid fall of temperature, to be followed in each experiment with the thermometer, ending in death. As one or a few colonies vigorously developed on agar can furnish a quantity of the cultivation-substance sufficient for the animal-experiment, the great advantage of preparing agar-cultivations at the earliest possible moment is evident. Great value must be attached to the animal-experiment as well as to the cholera-red reaction, because it demonstrates a quality belonging exclusively to the cholera-bacteria in a comparatively short time. Among all curved, that is, spirilla-like bacteria which have to be taken into account in cholera-diagnosis, none has yet been found which, if given in the dose stated, produces symptoms even approximately resembling those produced by the cholera-bacteria.

The six methods described above afford us the means of arriving at a sure and speedy diagnosis in all cases. All that is necessary is to apply them rightly, that is, in judicious sequence and combination.

The formerly used cultivations in the hollow object-glass and on potatoes, and the needle-point-cultivation in gelatine, have been rendered superfluous by the aids here described.

Supposing the proper methods of taking, packing, and

sending the objects of investigation to be known, the course of the investigation proper would be as follows :—

1. In the first place microscopic preparations of mucus-flakes, if possible, are made and examined. If the above described characteristic arrangement of the cholera-bacteria, or a pure cultivation of the same, appears in them, which, as has been said, may happen in nearly half of the cases, one declares the disease in question to be Asiatic cholera. For the complete verification of this diagnosis, a gelatine-plate-cultivation and a peptone-cultivation are prepared simultaneously. The former is placed in an incubator set at 22 degrees centigrade, the latter in one set at 37; there must therefore always be two incubators with different temperatures at disposal. After about eight hours cholera-bacteria in pure cultivation are found on the surface of the peptone-cultivation, in which case the indol-reaction is made; the gelatine plates show the characteristic forms of the cholera-colonies after about twenty hours. It has been already mentioned that in the Institute for Infectious Diseases this supplementary testing has invariably confirmed the provisional diagnosis in such cases.

2. If the microscopic investigation yields no sure result, gelatine-plate-, peptone-, and, if possible, agar-plate cultivations from the material under investigation must be prepared at once. The gelatine plates are kept at a temperature of 22 degrees centigrade, the agar plates and peptone-cultivation at 37. From six hours after the beginning of the investigation onward the peptone-cultivations are microscopically examined from time to time for curved bacteria. As soon as the latter appear agar plates are prepared anew, this time from the peptone-cultivation, in order to get large cholera-colonies as quickly as possible. This may be the case so early as ten hours after the beginning of the investigation; and one is then in a position, if their number is very large, and if meanwhile the gelatine plates also have developed cholera-colonies in preponderant majority, to pronounce the disease in question to be cholera, which diagnosis is afterwards tested by some

peptone-cultivations (prepared from the agar or gelatine plate) and the indol-reactions made with them.

3. It may happen, however, that comma-bacilli appear in the peptone-cultivation, though somewhat later and in smaller number; whereas at the same time no characteristic colonies, or only quite isolated ones, are to be found on the gelatine plates. Then all depends on the right using of the agar plates which were inoculated from the peptone-cultivations. On these, even under such difficult circumstances, suspicious colonies may possibly still appear. These must at once be propagated in pure cultivations on fresh agar plates, in peptone-tubes, and in gelatine plates, and used as soon as possible for the indol-reaction and experiments on animals; for, in such doubtful cases, it will always be necessary to make the animal-experiment too, in order to avoid the possibility of error. It may also be advantageous at once to cultivate in peptone-solution a second generation from the first peptone-cultivations which still show few comma-bacteria under the microscope, in order to obtain a further multiplication of the cholera-bacteria that may be present, and thus considerably facilitate the further investigation.

Even in such difficult cases, which, for the rest, are not frequent, the investigation can be finished in two days at most.

As, during the last epidemic, reports came several times of cases in which the cholera-bacteria were found only transiently in repeated investigations, it will be necessary in future under certain circumstances not to rest satisfied with a single negative result, but to repeat the investigation.

The investigation of water for cholera-bacteria demands special discussion. It was made many times in the course of last year's epidemic, but, except by C. Fränkel[*] and Lubarsch,[†] always with a negative result, even at places

[*] German Medical Weekly, 1892, No. 41.   [†] *Ibid.*, 1892, No. 43.

where the presence of cholera-bacteria in the water was to be confidently expected. The failure of the demonstration has, of course, been turned to account in antibacterial quarters against the importance of the cholera-bacteria. All previous experience, however, teaches that man is the finest reagent on cholera-bacteria, and the occurrence of cholera-cases among human beings who could be infected only by means of water is in itself an irrefutable proof that cholera-bacteria must have been in the water in question. Such occurrences were frequently observed in the last epidemic, and especially the course of the epidemic in Hamburg, Altona, and Wandsbeck showed in an indisputable manner that water may be the bearer of the contagious matter of cholera. From the vain attempts hitherto made to find the cholera-bacteria in water I have therefore not concluded that they are not in it, but that our method of investigation is too imperfect to detect them under the very difficult circumstances with which the bacteriological demonstration of contagious substances in water has to contend.

The main difficulty lies in the circumstance that water contains more or less numerous other bacteria, which at once overgrow and stifle the cholera-bacteria in artificial cultivations. This is especially true in the case of much polluted waters, into which the dirty water of towns and cities, fæces, etc., flow, and these are just the waters that had to be examined for cholera-bacteria. For this reason one could hope to find something with the means formerly at disposal, only if the water was strongly diluted before the investigation, in order somewhat to eliminate the influence of the other bacteria. Of course, the dilution of the water had another consequence, namely, that only very small quantities, generally only fractions of a drop, could be examined. Under these circumstances success could be counted upon, only if the number of the cholera-bacteria in the water was extraordinarily great, or if that of the other bacteria was very small, conditions which depend on accident and are perhaps quite exceptional. It is probably due to such fortuitous circumstances that I succeeded in demonstrating the presence of cholera-bacteria in an Indian tank, Fränkel in the water of Dortmund harbour, and Lubarsch in the bilge-water of an Elbe steamer. In the two former cases the investigation seems to have been aided by the large number of the cholera-bacteria, in the latter by the small number of the water-bacteria.

In order to be no longer dependent on such accidental circumstances, nothing was left but to cast about for a better method of investigation, and, after the favourable experience which had been made with the enrichment of cholera-fluids by the peptone-cultivation, nothing was more natural than to apply the same principle to the investigation of water.*

At first, just as in the peptone-method described above, one drop or a few drops of the water to be examined were added to the peptone-solution. As, however, it was to be inferred from the former futile investigations of water that it is quite exeptional to find very numerous cholera-bacilli in water, the method was modified so that the largest possible quantities of water were subjected to investigation, which was done by directly adding a sufficient quantity of peptone and common salt (1 per cent. of each) to the water, and then keeping the mixture at 37 degrees centigrade.

After ten, fifteen, and twenty hours agar plates must be inoculated from the peptone-cultivation. The microscopic examination of the peptone-cultivation is of subordinate importance in this case, as one cultivates curved bacteria, very like cholera-bacteria in form, out of almost every water in the manner stated. On the other hand, all colonies of suspicious appearance that develop on the agar plate are first microscopically examined, and, if they consist of curved bacteria, further cultivated in order to be tested by the indol-

* Probably following the same line of thought, van Ermengens, Bujwid, and Arens have proposed methods similar to that described above. With the help of his method van Ermengens has succeeded in demonstrating the presence of cholera-bacteria in a water-course on the banks of which cholera was prevalent. Compare van Hassel's Rapport sur l'épidémie de choléra 1892 dans la commune de Paturages.

reaction and the animal-experiment, which must under all circumstances complete the diagnosis in water-investigations. According to the experience hitherto made it seems to be advisable not to make the quantity of water to be examined greater than about 100 cubic centimetres, and rather to investigate a number of specimens than to go beyond this quantity.

The method just described has stood the test of experience perfectly in the Institute for Infectious Diseases; the presence of cholera-bacteria was demonstrated with its aid in the water of the Elbe, in a well in Altona, on the sewage-fields of Nietleben, in the water of the Saale, and in the water taken from the conduit-pipes of the Nietleben asylum during the winter epidemic in Hamburg, Altona, and Nietleben.* In these cases, in which the detection of cholera-bacteria in a considerable number of suspicious specimens of water succeeded for the first time, all care was of course taken to exclude the possibility of mistake. Most of the investigations were made simultaneously by several of the gentlemen employed in the Institute. For comparison, other specimens of water not suspected of containing cholera-bacteria, and specimens taken from the suspected places repeatedly after as well as before the cessation of the epidemic, were examined. The various investigators almost always arrived at harmonious results, the micro-organisms always identified as cholera-bacteria by the indol-reaction and experiments on animals were found only in waters connected with cases of cholera, and the cholera-bacteria disappeared when the epidemic ceased.

As already mentioned, curved bacteria which accumulate just as the cholera-bacteria do in the upper layers of the peptone-cultivation are found almost regularly, at least very frequently, in water of the most various origin. Almost a dozen such bacteria, belonging to the spirilla, are already collected in the Institute for Infectious Diseases. Other bacteriologists also have found such spirilla in water. All of

---

* I reserve a more exact account of the circumstances under which these results were obtained for another communication.

these can be more or less distinguished from the cholera-bacteria by a practised eye even by the appearance of the agar and gelatine-colonies, but they can be distinguished with great facility and certainty by the absence of the in-dol-reaction and of the poisonous effect on guinea-pigs. In human excreta they seem to occur extremely seldom and never in large quantity, so that there is no danger of their adding to the difficulties of the bacteriological diagnosis of cholera.

Even the method formerly applied in the bacteriological diagnosis of cholera, which was restricted to gelatine-plate-cultivation, puncture-cultivation, etc., demanded no small amount of practice, if its purpose was to be quickly and surely gained. The somewhat more complicated new procedure, however, demands manifold practice and complete mastery of the technics in a still higher degree, and it is therefore certainly not superfluous if I point out once more in conclusion with special emphasis that everyone who has to make the bacteriological diagnosis of cholera should acquire the necessary practice in time, and that everyone who does not possess such practice would do better to relegate the investigation to competent hands.

Water-Filtration and Cholera.

(*Received by the Periodical for Hygiene and Infectious Diseases on the 26th of May 1893.*)

By all investigators of cholera whose judgment was not disturbed by fantastic dreams about telluric-cosmic influences, or bound by obstinate adherence to theories refuted long ago, more or less great importance has from of old been attached to water as a bearer of the contagious substance of cholera. Only as to the greatness of the influence of water did opinions differ. Some, misled by isolated observations of a specially surprising nature, have gone decidedly too far in this direction, and declared water to be the exclusive bearer of the germ of cholera. To these the designation "Water-Fanatic" or "Water-Theorist," which has been much used of late, may not quite unjustly be applied. That I do not belong to this class of cholera-investigators will be admitted at once by everyone who knows what I have said and written about cholera. I have always said that, according to the experience hitherto gained, direct transmission from person to person is possible but to all appearance not very frequent, that, on the other hand, indirect transmissions by many bearers of the germ of cholera play the principal part in the real epidemics and mass-outbreaks of that disease, and that among these bearers water is one of the most important. I have further endeavoured to show by examples that, under certain conditions, water has really played the part ascribed to it. Beyond that, however, I have never, so far as I know, expressed an opinion as to the extent to which this factor is to be regarded as effective. Nor has it been possible hitherto to arrive at a definite judgment on this subject, because investigations of the relations of cholera to water have almost always been undertaken from a onesided point of view, and are therefore generally open to objection. Why, under such circumstances, attempts have been made to stamp me of all people as a "Drinking-Water-Fanatic," I do not clearly understand. It seems almost as if the purpose were to attribute to me by hook or by crook opinions which it is easy to refute.

In the last epidemic indeed, as, I suppose, nobody will seriously dispute, water played a very important part. Nevertheless we cannot yet know even now whether that will be the case in future too, and it is certainly wiser not to pronounce a definitive opinion on the importance of water till we have gathered sufficient experience. Last year, however, has at any rate shown once more that we have every reason to devote the greatest attention to water-supply.

The cholera-epidemic in Hamburg, Altona, and Wandsbeck was extremely instructive in this respect. These three places, which immediately border on one another, and, strictly speaking, form only one single city, do not essentially differ from one another except in the one respect that they are differently supplied with water. Wandsbeck receives filtered water from a lake scarcely exposed at all to pollution with fæces, Hamburg takes its water in an unfiltered state from the Elbe at a point above the city, while Altona gets its supply in a filtered state from the Elbe below the city. Hamburg, as is well known, suffered terribly from cholera, whereas Wandsbeck and Altona, if the cases introduced from Hamburg be deducted, hardly suffered at all. The most surprising phenomena of the epidemic came to light on the frontier of Hamburg and Altona. On both sides of the frontier the state of the soil, the buildings, the sewerage, the population, in short all the conditions that are of importance in this connection, are perfectly similar, and yet the cholera in Hamburg spread only to the frontier of Altona, and stopped there. In one street, which forms the frontier for a considerable distance, the Hamburg side was attacked by cholera, while the Altona side remained free of it. In one group of houses, the Hamburger Platz, the cholera even did more than a man could have done with the best maps of the frontier between Hamburg and Altona at his disposal. It found out with sharp precision not the political frontier but the frontier of the water-supplies of the two cities. The said group of houses, densely peopled by workmen's families, belongs to Hamburg, but is supplied with water by Altona,* and it remained absolutely free of cholera, whereas numerous cases and deaths occurred in the Hamburg houses all round. Here then we have to do with a kind of experiment, which performed itself on more than a hundred thousand human beings, but which, despite its vast dimensions, fulfilled all the conditions one requires of an exact and absolutely conclusive laboratory-experiment. In two large groups of population all the factors are the same except one, namely, the water-supply. The group supplied with unfiltered Elbe water suffers severely from cholera, that supplied with filtered water very slightly. This fact gains in significance when we consider that the Hamburg water is taken from a place where the Elbe is still comparatively little polluted, whereas Altona has to use the Elbe water after it has received all the fluid refuse, including the fæces, of nearly 800,000 people. Under such circumstances there is at first sight for all who are accustomed to reflect on the facts of natural science no other explanation whatever than that the difference which the two groups of population show as regards cholera is due to the difference of the water-supply, and that Altona was protected against cholera by the filtration of the Elbe water. It is impossible simply to deny this fact, and all that remains is to

---

* Besides this group of houses there is only one building in Hamburg, a brewery in Sankt Pauli, which receives water from Altona. I am able to state most emphatically, on the basis of information received by myself in Altona and Hamburg, that the assertions that other Hamburg houses are connected with the Altona water-conduits and had cholera nevertheless, and that, on the other hand, there are Altona houses that get their water from Hamburg and yet had no cholera, are utterly incorrect.

try to bring it into harmony with one's views of the nature of cholera. As we have to do in this case with an epidemiological fact of the first rank, which lies before us in perfect clearness and transparency, which also, owing to its easy accessibility, can be further tested and supplemented in all directions as regards the correctness of the observations on which it is based, and is therefore in this respect simply unique of its kind, one has a right to demand that every investigator of cholera who claims consideration for his opinion shall tell us what he makes of it.

For the bacteriologist nothing is easier than to give an explanation of the restriction of the cholera to the sphere of the Hamburg water-supply. He need only point out that cholera-bacteria got into the Hamburg water either from the outlets of the Hamburg sewers, or, which is much more likely, from the excreta of cholera-patients on board the numerous Elbe barges anchoring off the place where the water is taken from the river, and that, after this had happened, cases of cholera, more or less numerous according to the degree of pollution, could not but occur among the people who used that water. The town of Wandsbeck was spared, because its water was not exposed to such pollution, and was also filtered. Altona received water which was originally much worse than that of Hamburg, but which was wholly or almost wholly freed of cholera-bacteria by careful filtration. This view is in perfect harmony with all bacteriological experience, and with our present knowledge of infectious substances; it contains no contradictions and nothing forced or artificial.

How people would derive the phenomena of the Hamburg-Altona cholera-epidemic from cosmic-telluric or from purely meteorological factors, I am at a loss to imagine; for sky, sun, wind, rain, etc., were distributed with absolute equality on both sides of the frontier. I hardly believe that the adherents of the cosmic-telluric theory will even attempt to find an explanation.

I was particularly curious to know what the soil-theorists or localists, as they now like to call themselves, would say to the Hamburg-Altona epidemic. According to what has hitherto transpired on the subject, they now, though after some resistance, admit the fact that the cholera-epidemic proper remained restricted to the territory which received unfiltered Elbe water, and that therefore the unfiltered water must have exercised an influence; but the explanation of the way in which this influence, with the co-operation of the soil of course, was brought to bear, has been given in so extremely inadequate a manner that I should hesitate to discuss it here, if it had not come from so highly respected a quarter.

The localistic view [*] is that the water had not an infecting but a "predisposing" effect, the unfiltered water having brought dirt into the dwellings, the streets, and the soil, and thus in a manner created a suitable nutritive medium for the development of the cholera-germ. In putting forward this theory the revered Herr Localist in his embarrassment probably did not reflect how infinitesimal, even in the most unfavourable cases, the quantity of dirt is that can be deposited in dwellings and in the soil by water used for cleansing purposes, compared with the infinitely larger masses of dirt which human housekeeping daily brings into the dwellings, and which are constantly deposited by human beings and animals on the streets and courtyards. He has also left out of account that the Elbe water which gets into the conduit-pipes does not always contain an admixture of the sewage of the city of Hamburg, but only quite exceptionally, when the tide is unusually high. He seems, however,—and this is the heaviest reproach to be brought against him here,—to have quite forgotten that Hamburg is one of the best drained cities in the world, that is, is furnished with works of which we know that they convey the dirty water out of the houses and away from the courtyards and streets by the shortest way to beyond the limits of the city. Of what earthly use is the

[*] Von Pettenkofer, "On Cholera and the Last Cholera-Epidemic in Hamburg." Reports of the meetings of the Medical Society of Munich, 1892, vol. ii. Von Pettenkofer, "On the Hamburg Cholera-Epidemic of 1892, and on Protective Measures," Munich, 1893.

sewerage, if it cannot even remove and render harmless the small addition of organic matter which unfiltered water brings with it? The soil-theory could not possibly give a more signal proof of its utter failure than this unlucky attempt at an explanation affords.

At any rate the localists too will have in future to regard the filtration of water not as a thing of mere secondary importance, but as a very useful and even, under certain circumstances, indispensable measure; and one sees here once more, as in so many other cases, that people may differ as to the explanation of a matter, but arrive at the same results as regards its practical treatment.

The Hamburg-Altona cholera-epidemic, then, irrefutably proved that the filtration of water through sand, in the manner in which it is effected in Altona, affords a practically sufficient protection against cholera-contagion. I expressly emphasise the fact that, if filtration is to afford protection, it must be managed as in Altona. I have inspected a considerable number of waterworks with filtering apparatus, and know that but few of them fulfil the prescriptions now valid as regards filtration so strictly as is done in Altona, and I have every reason to believe that cholera would not everywhere have been warded off so successfully as there.

In order to make myself intelligible to readers to whom the technics of filtration are not sufficiently known, I must go a little into the details of sand-filtration before proceeding further.

The purpose of filtration is to cleanse water of the substances suspended in it. Dissolved matter goes through the filter quite unaltered or hardly perceptibly altered. As the chemical investigation of water busies itself in the main with the examination of its dissolved ingredients, it could contribute nothing to the study of the processes of filtration. But people used to be so accustomed to judge of the quality of water by its chemical properties that many attempts have been made, in utter misconception of the circumstances of the case, to follow and to test the process of filtration chemically. Such attempts of course never produced any useful results. The technicians of filtration soon recognised this fact, and tried to help themselves in another way. They tested the transparency of the water in glass or metal cylinders, the so-called water-testers, before and after filtration, and judged of the efficiency of the sand-filters accordingly. By this simple means they succeeded in discovering the most important conditions of the adequate cleansing of water from suspended ingredients. They found out, namely, that the filtration proper does not take place in the sand itself, but that a layer of mud must first form on the surface of the sand by deposits from the still uncleansed water, and that this layer of mud is the real filter which keeps back the dirt suspended in the unfiltered water. In the management of filters therefore the essential thing is that the filtering mud layer regularly form first, that it be not disturbed during filtration, and that, after constant further deposits of dirt have made it too thick and too little permeable by the water, it be removed at the right time. To all appearance different waters differ very much, according to the quantity of mineral and vegetable matter they hold in suspension, in capacity for furnishing the filtering mud layer. Some river-waters which are specially rich in clay can deposit a good filtering layer in eight to ten hours. Other kinds of water, the turbidity of which is caused more by vegetable matter, need twenty-four hours at least. At certain times of the year, when the water "blooms," the vegetable ingredients suspended in it increase to an extraordinary extent owing to the presence of masses of microscopic algæ in the water; they are at the same time of an unusually slimy nature, and consequently form a mud layer which often becomes impermeable by water after only a few days, and must then be removed. Even from these brief indications it is evident that sand-filtration is by no means so simple a matter as is very frequently supposed. But it has been further ascertained that the sand layer, which is gradually wasted away by use, must not be allowed to diminish below a certain depth,—about thirty centimetres,—

and that the water must permeate the sand layer at a certain rate, about a hundred millimetres an hour, if the cleansing is to be as complete as possible. Some other less important rules for filtration need not be mentioned here; those above given are the most important, and suffice to render what follows intelligible.

If any disturbance take place in the working of the filter, and the above-mentioned rules cannot be observed in consequence, this shows itself at once in the quality of the filtered water. It no longer presents a perfectly transparent appearance on examination with the water-testers. To the consumer, indeed, this is hardly ever perceptible, for even such inadequately cleansed water may still look clear in carafes and tumblers. Partly for this reason, partly because the disturbances in the working of the filter are generally of but short duration, no great importance used to be attached to them, and in many cases it was not even deemed necessary to construct waterworks in such a manner that the insufficiently cleansed water could be excluded from consumption.

A complete change has taken place, however, since we gained a better knowledge of contagious substances, and arrived at the conviction that water must not only be rendered clear by filtration, but must also above all be freed of infectious substances that may have got into it. For the detection of infectious matter in water, the water-testers were of course inadequate; the only possible method was bacteriological investigation, and since that time it has been recognised on all hands as the only reliable means of testing the adequacy of the cleansing of water by filtration. The application of bacteriological investigation to the study of the processes of filtration has in general confirmed what was known before, but it very soon showed that even slight and rapidly transitory disturbances in the working were by no means to be made so light of as had formerly been done. If a filter is working satisfactorily in every respect, it is found by experience that the filtered water contains fewer than a hundred germs capable of development per cubic centimetre, no matter whether it contained several hundred thousand or only a few hundred per cubic centimetre before filtration.* But the slightest disturbances of the process of filtration—for instance, the increase of the velocity beyond 100 millimetres, injuries to the mud layer, etc.—are at once followed by an increase in the number of germs in the filtered water. The small number of germs still remaining when the filter is in good working order undoubtedly emanate for by far the greater part from the filtering material covered with bacterial vegetations, that is, from the lower parts of the sand layer, the gravel and stone layers, which, like all water-washed objects, get overgrown in course of time with vegetations of micro-organisms. These are of course quite harmless, and are found in every kind of water. To judge from the investigations of Fränkel and Piefke,† however, a small number of the bacteria found in filtered water come from the unfiltered water. Even with our best filtering arrangements then we cannot keep back all micro-organisms. This too must be attainable, but it would then be necessary to make the rate of filtration much slower even than it is, and perhaps to use thicker layers and other filtering material, all which would involve an enlargement of the works and an increase of expense that would exceed practicable limits. To all appearance we have attained the limit of capability with our present arrangements.

Altona's experience of last year, however, is the best proof that the efficiency already attained is practically adequate. At the waterwork of that city, which is fortunately managed by one of our ablest and most experienced filtration-technicians, the rate of 100 millimetres was never exceeded during the cholera-epidemic, and it was evidently due to this

* When filtration goes on at the rate of 100 millimetres an hour, differences in the bacterial content of the water before filtration may indeed affect the number of the bacteria in the filtered water, but the latter number must never exceed 100 germs per cubic centimetre, even when the unfiltered water is very strongly polluted.

† Periodical for Hygiene and Infectious Diseases, vol. viii., No. i.

precaution that Altona suffered in so strikingly slight a degree. According to the information at my disposal regarding the cholera in Altona,* there were about 500 cases there, of which, however, at least 400 were found to have been introduced from Hamburg, or to have originated through intercourse with the harbour, or must be regarded as emanating from such cases by direct transmission. Of the remaining 100 cases, which cannot be directly or indirectly traced to Hamburg, a certain number will nevertheless have been of the same origin, though it was not possible to prove it. Whether among the remnant of cases left after these deductions, which, as having originated independently of Hamburg, form the Altona epidemic proper, there were a certain number of cases which were due to the imperfect cleansing of the infected Elbe water in the sand filters, can, I suppose, never be decided. After the investigations of Fränkel and Piefke,-however, which have shown that, at a filtration-rate of 100 millimetres, the cholera-bacteria are not completely kept back by sand filters, I think it very probable. But the number of cholera-cases that originated in Altona in this manner cannot have been at all considerable, and it will probably not be possible to demand that, in order to prevent such isolated cases with certainty in future, the sand filters must be enlarged to two or even three times their present dimensions. All that we yet know of sand-filtration, therefore, compels us to admit that, even under the most favourable circumstances, it cannot afford absolute protection against infection, though, as I have already said, it does afford a protection with which, considering the practical conditions of life, one may rest content.

In itself this result will afford us no small reassurance with regard to all towns that have to trust to sand-filtration. But the present cholera-epidemic itself has unfortunately already taken care not to let this reassurance go too far; it has shown us by several palpable examples that filtration-

* Compare Wallichs "On the Cholera in Altona," German Medical Weekly, 1892, No. 46.

apparatus in itself does not suffice, but that it must be perfectly constructed in every respect, most carefully managed, and kept under constant bacteriological supervision, if the protection afforded is not to be more or less illusory. The occurrences which have again thrown us into some uncertainty with regard to the efficiency of sand-filtration are those of the cholera-epidemic at Nietleben near Halle and of the winter epidemic in Altonà. They are so instructive that they deserve somewhat minute discussion.

The provincial lunatic-asylum of Nietleben, situated some kilometres west of Halle, gets its water from an arm of the Saale which is called "the Wild Saale." The water of the Wild Saale flows down into a filtering apparatus and then through an iron siphon-pipe to the pumping station of the establishment. The pumps then lift it into several highly situated reservoirs, from which the pipes are fed. The filtering apparatus, to which our interest is at present limited, has been at work since 1883. In its arrangements it almost entirely resembles the larger filter-works built in recent times. It consists of three walled and covered filters side by side, each having an area of 41·25 square metres. The filtering material consists, as usual, of an undermost stone layer, on which rest first gravel and then sand, the latter in a layer 66 centimetres deep. The river-water does not flow directly on to the filters, but passes first through two parallel basins with an area of 11·25 square metres each, from which it is led through iron pipes to the surface of the sand layer. After the water has passed through the filter-layers, it gathers in channels at the bottom of the filters connected with three smaller receivers, the so-called collecting shafts, on the other side of the filters. These collecting shafts, which communicate with one another, form, as it were, the pure-water-reservoir, from which the iron siphon-pipe of the pumping station sucks the filtered water. According to the evidence of the man in charge of the filters, which was protocolled, the filters were treated as follows :—

As soon as one of the filters was so choked with mud that

it had to be cleaned, its inlet and outlet were closed, and the water standing on the sand was pumped off with a handpump till it was only 10 centimetres deep. Then the filter was reconnected with its collecting shaft, and the water in the latter was removed by means of pumps, after the communication with the siphon-pipe had been closed. The water left standing on the sand layer of the filter now sank in the filtering material, leaving the surface of the sand dry. Then boards were laid on the sand, and all of the uppermost sand layer that looked dark and muddy was removed. After this the surface of the sand was smoothed again, and the boards, which had served as a basis for the workmen's feet, were removed. In order to fill the filter again, the slide which shut off the water of the fore-basin from the filter was opened very gradually, in order that the upper sand layer might not be whirled up, which, however, could not be quite avoided in the parts near the pipe, even when the water flowed in slowly. The opening of the slide lasted about twenty minutes. About three quarters of an hour after the slide had been opened the water generally stood higher in the collecting shaft than in the two other collecting shafts, whereupon the connection with them and with the siphon-pipe was re-established, and the process of filtration recommenced.

This filtering apparatus was originally estimated for 700 inmates of the establishment at the rate of 300 litres daily per head, the pumps working 15 hours a day. Consequently 14 cubic metres of water had to be filtered per hour. For this only two of the filters, with an area of 82·5 square metres, were at disposal, as the third, lying idle while the work of cleaning it was going on, was not taken into account. Calculating with these numbers for a basis, one gets a filtration-rate of 170 millimetres an hour.

The filters remained efficient for about three weeks after the work of filtration first began. Very soon, however, (already in 1884) the filtration became less satisfactory, and the filters had to be cleaned at intervals varying from a fortnight to ten days. It was found at the same time that the siphon-pipe, which, as already mentioned, conveyed the filtered water to the pumping apparatus, and originally consisted of clay pipes, was not impervious. It received pretty considerable quantities of underground water from the ground in which it was laid. In reality, then, the filtering apparatus even at that time, when the filters were cleaned every ten to fourteen days, only partly supplied the wants of the establishment. The clay siphon-pipe was then replaced by an iron one, so that only filtered Saale water could get into the asylum. At the same time the establishment was considerably enlarged, and the number of the inmates increased from 700 to about 1000. In this way, however, the disproportion between the capability of the filtering apparatus and the demands made on it, which had existed from the very first, was so tremendously increased that, in order to obtain the necessary quantity of water, only one expedient remained, namely, that of keeping the sand layers of the filters as permeable as possible. The intervals between the cleanings became shorter and shorter, and in the last two years (1891 and 1892) the man in charge had to clean two of them every week in summer and all three in winter. Number one was cleaned every Monday and Tuesday, number two every Wednesday and Thursday, number three every Friday and Saturday; on Sundays all three worked together.

The filtering apparatus has some faults even of construction. It has no contrivance for measuring and regulating the rate of filtration; badly filtered water cannot be removed; in starting the filter the sand layer cannot be filled with filtered water from below, and the wallowing up of the sand by the instreaming unfiltered water cannot be prevented. These are faults, however, which are found in large filter-works too. They make it very difficult to get well filtered water, but not impossible. Had the Nietleben filter-work been skilfully managed, and kept under constant bacteriological supervision, it would have furnished water less in quantity indeed,* but of

* With two of the filters in full activity, and working with a total filter-area of 82·5 square metres at the rate of 100 millimetres an hour, the work would have

a quality hardly open to reproach from the hygienic point of view. After the above description, however, it is impossible to speak of the treatment of the filters as skilful. It is evident that none of the parties concerned had the least conception of the manner in which a filter-work must be treated. The filtered water was chemically examined occasionally, and it was thought that, if the chemical analysis showed it to be free of ammonia and nitrous acid, the filter-work was doing its duty, especially if the chemist at the same time reported that there were indeed some brown and green algæ visible in the water under the microscope, but that no "cocci and bacteriaceæ" ("Kokken und Bakteriaceen"), and "no things that could be called exactly noxious," were to be found in it. Not one regular bacteriological investigation was made before the outbreak of the cholera; and yet a single such investigation would have shown at once that the water was freed in the filter-work only of the coarser substances suspended in it.* Of micro-organisms and of infectious substances in particular only the smallest part could be kept back by such filtration. At bottom the one thing aimed at was to hurry as much water as possible through the filter-beds, and, with this aim in view, it was utterly impossible to proceed in the regular way, and to let the indispensable mud layer form on the surface of the sand.

The fact that the water must rest at least 24 hours over the sand, in order to let the filtering mud layer form, was utterly unknown at Nietleben, for from the first the filtration was started three quarters of an hour after the admission of

furnished 198 cubic metres of filtered water, that is, nearly 200 litres a head daily, provided that the pumping apparatus had worked 24 instead of 15 hours a day. If necessary, however, the demands made on the work might still have been much reduced, for 200 litres a head are a very copious water-supply. I had occasion lately to inspect another lunatic-asylum, which, indeed, was not yet in use, but of which I was officially assured that it would have to make a water-supply of 25 litres per head daily suffice, as it was absolutely impossible to procure more. It is true that it has no steam-boilers to supply with water, as the Nietleben asylum has.

* Not always even of these, for the water in the asylum was often so unclear that the employees did not like to drink it.

the water to the filter. But so much mud must nevertheless have been deposited on the surface of the sand during filtration after one day or a few days that the water would then at least have been well filtered. In the last years, however, even this was no longer allowed, for the filters worked only about five days at a time, and the filtering mud layer was scratched off again almost before it had formed. At bottom therefore in these years the filtration was merely nominal, in reality the water went uncleansed through the filter-work. In order to prove this, Professor Pfuhl bacteriologically examined the water before and after filtration, while the process of filtration was going on. In the uncleansed Saale water he found 302,400 germs per cubic centimetre, in the filtered water 52,410. Had, as I have already said, only one such bacteriological investigation been made before the cholera-time, and its result duly appreciated, the danger could hardly have been looked forward to so carelessly; the evils would have been remedied in time, and the establishment would have been saved from a great disaster. And there was every reason to pay the very greatest attention to the filtering apparatus, for, as I shall have to show at another place, the circumstances at Nietleben are quite similar to those at Altona. The water to be filtered is most alarmingly exposed to pollution. The Saale water flows to the place laden with the dirt of many places situated farther up, especially of Leipsic, the sewage of which is poured into the Saale by the Elster, which flows into it four to five miles above Nietleben. The worst circumstance, however, is that a dirty little brook, bearing the very significant name of Saugraben (Sowditch), flows into the Saale, fifty paces above the place from which the asylum takes its water. It comes from the village of Nietleben, receives the dirt of that place, of several mines and factories, and, immediately before flowing into the Saale, the off-flow of the sewage-fields of the asylum itself. Nor does the turbid water of the Saugraben mingle quickly and sufficiently with the Saale water; on the contrary it flows for a long time, recognisable by its colour, along the bank, and arrives almost undiluted at the place where the asylum takes its water, so that it is not pure Saale water but only the moderately diluted water of the Saugraben that flows into the filters. The danger of constructing a waterwork close below so strongly polluted an affluent would certainly not have been risked, unless absolute confidence had been felt that the water would be cleansed of noxious substances by careful filtration; and it was certainly not suspected that, managed as the waterwork was, it would not afford the asylum protection against infection, but must necessarily become simply a catching-apparatus for infectious matter, which accordingly, sad to say, it ultimately became. The details of the cholera-epidemic at Nietleben were in themselves a proof that the badly filtered water was the bearer of the infectious substance. Moreover cholera-bacteria were found in the water of the Saale at two different times below the place where the Saugraben flows into it. They were also found in filtered water taken from the collecting shaft of filter number 2, and in water taken from the pipes in the asylum itself.

In the waterwork of Nietleben we have an instance of an establishment which, despite its faults, must have yielded satisfactory results, if it had been very skilfully and carefully managed; by means of slight alterations too the faults might have been almost wholly remedied. In contrast with this, however, we are now to learn by the example of the Altona waterwork that the filter-works now existing may have faults, the importance of which has either been unknown hitherto or at least inadequately appreciated, and against which even the best management is powerless.

The Altona waterwork is one of the oldest in Germany. It is at Blankenese, about seven miles below the place where, on the same bank of the river, at the frontier of Hamburg and Altona, all the sewers of those two cities pour their contents into the Elbe. The place from which the water is taken lies in the part of the river subject to the tide, the average difference in the height of which in the streams of that region is somewhat more than two metres.* During the ebb the waterwork receives the Elbe water of the right bank, strongly polluted by the sewage of the two cities; during the flow it receives from the backrolled stream water containing large quantities of less polluted Elbe water. This difference can be perceived at once by the bacteriological examination of the Elbe water, the germ-contents of which vary very considerably; at short intervals a few thousand and then again some hundreds of thousands of germs can be found in the cubic centimetre, and that even in winter, when the number of germs in the water of other rivers is wont to diminish considerably and uniformly. Close to the Elbe is the pump-work, which raises the water 84·7 metres high to the Baursberg (a hill overtopping the steep bank of the Elbe), where the waterwork is built. The latter consists of two settling-tanks, in which the pretty turbid Elbe water deposits a part of its heavier ingredients, and ten *open* filters. Each filter has an area of somewhat more than 800 square metres. From the filters the water flows into a pure-water-reservoir containing 3050 cubic metres, and from there down to the city. The construction of the filters does not differ from that usual in works of the kind.

Since the summer of 1890 the water has been subjected to bacteriological examination, but only once a week, and only in specimens taken from the pure-water-reservoir.† Till summer 1892, that is, during a period of two years, except for a short time in January 1891, the number of germs in the cubic centimetre of filtered water has always remained less than a hundred. Numbers under twenty were the rule; fifty to seventy colonies were exceptional. This excellent result is undoubtedly due to the skilful and careful treatment of the filters; specially strict care was taken that the rate of filtration never exceeded 100 millimetres an hour;

* Compare "Hamburg from the Points of View of the Naturalist and the Medical Man" (a treatise written in honour of the forty-ninth assembly of German naturalists and medical men), Hamburg, 1876, page 236.

† Kümmel's "Experiments and Observations on the Effects of Sand Filters," Schilling's Journal for Gas-Lighting and Water-Supply, 1893, No. 9.

generally it was still slower. It is of special importance too that the uniformly low numbers of germs were maintained, though the unfiltered water kept constantly fluctuating in this respect, and was sometimes polluted to an extraordinary degree. This is a proof that, with proper management, a filter-work must furnish water with fewer than 100 germs per cubic centimetre, even when the unfiltered water is of the worst, for it would be hard, I think, to find worse water than that of the Elbe at Blankenese.

But we can also learn from the Altona filter-work that disturbances in the working of the filters at once reveal themselves bacteriologically; not rarely, however, also by the effect of the infectious substances no longer sufficiently kept back by the filter, that is, by the outbreak of infectious diseases in the districts supplied by the waterwork. Fortunately there are not many infectious diseases the germs of which can be carried by water. We know only two with certainty,—cholera and enteric fever. For us cholera comes under consideration only periodically, but enteric fever, as an endemic disease occurring everywhere among us, constantly. One may assume, I think, that our cities are never quite free of enteric fever, and that their sewage always contains, besides other fæces, those of enteric-fever-patients also, and in them the infectious matter of that disease. If now, as is generally the case nowadays, the sewage is removed as soon as possible from the place by the sewers, and if it is not brought in again by the water-supply, enteric fever, at least so far as infection by water is concerned, abates. It need not on that account wholly disappear, for, like cholera, it has other ways of spreading than by water; I even believe that enteric fever is still less dependent on water, and can find its way more easily without water, than cholera with its exotic, strictly speaking, tropical infectious matter, the life of which depends on moisture and warmth in a much higher degree than that of the hardier enteric-fever-bacillus. The latter, however, can, as numerous recent observations prove, occasionally be spread by water,* and, if

---

* If I say here that enteric fever can be spread by water, I am quite prepared

a striking increase of equably spread cases of enteric fever without connection among themselves suddenly occurs in districts supplied by a certain waterwork, there is every reason to inquire whether the epidemic is not caused by the water-supply. If now, in such a case, this suspicion is confirmed, enteric fever has served, so to speak, as an index for defects in the water-supply. Such a connection has in point of fact proved demonstrable in almost all such cases in which the investigation was conducted thoroughly and without prejudice, and the phenomena of enteric fever in a place will therefore in future be a most valuable aid to the detection of faults in the water-supply.

Something of this kind happened in Altona. During the last decades the city has had epidemics of enteric fever which, owing to their peculiar phenomena, awakened the suspicion that they must be connected with the water-supply. We owe our first information on this subject to Reincke.* He pointed out that the enteric-fever-epidemics in Altona are not simultaneous with those in Hamburg, but some weeks subsequent to them, and that their sphere in Altona coincides with that of the water-supply. He therefore expressed the suspicion that they were due to the water-supply. He was confirmed in this suspicion by the result of bacteriological investigations carried out by Spiegelberg in 1885 and 1886. These investigations, which were made unfortunately only at intervals of a month, showed that the number of the germs in the water of the Altona water-supply had greatly increased shortly before the outbreak of the enteric-fever-epidemic of 1886. On the 20th of February 1886 the number was 1150 per cubic centimetre. Reincke concluded from this that disturbances must have occurred in the working of the filters, but did not succeed in demonstrating them. Wallichs† got a step nearer to the

to see myself at once denounced in a certain quarter as a "drinking-water-fanatic" as regards enteric fever too.

* Reincke's "Enteric Fever in Hamburg, with Special Reference to the Epidemics of 1885 to 1888," Hamburg, 1890, pages 35 and the following.

† Wallichs' "An Epidemic of Enteric Fever in Altona in the beginning of 1891," German Medical Weekly, 1891, No. 25.

solution of this question. It had specially struck him that from 1886 to 1888 an enteric-fever-epidemic had developed every year after a long frost. In 1886 a period of frost in February was followed by an enteric-fever-epidemic in March. In 1887 January was the colder month, and enteric fever rose to its maximum in February. In 1888 February was colder again, and enteric fever was severest in March. In 1891 also a cold period in January was followed by an increase of enteric fever in February. During this last epidemic the results of the above-mentioned continuous bacteriological investigations of the Altona water were at disposal. They showed, as Spiegelberg's investigations had shown in 1886, that shortly before the outbreak of the enteric-fever-epidemic the number of germs in the filtered water had risen considerably (to 2615 on the 13th and to 1364 on the 20th of January according to Wallichs, to 1900 once and to 1100 another time according to Kümmel). Wallichs, indeed, did not then feel justified in directly attributing the epidemic to the water-supply, but he expressed the strong suspicion that it might be the cause, and raised the question whether the filtration had not been disturbed by the forming of ice on the open filters; or whether it was not possible that, when the filters were cleaned during severe frost, the surface of the sand froze and then filtered inadequately. These possibilities were disputed by Kümmel at the time. The same writer afterwards opposed the theory of the dependence of the enteric-fever-epidemic in Altona on the water-supply, on the ground that another enteric-fever-epidemic broke out in February 1892, during and before which the number of bacteria in the filtered water did not increase. I do not admit the validity of this latter objection, for the filtered water was examined only once a week, and the water cleansed by each of the filters was not examined, but only the water of the pure-water-reservoir, consisting of water from all the filters. Under such circumstances, as we shall afterwards see, fluctuations in the bacterial content can easily be overlooked, especially if the disturbances in the filtration are restricted to only one of the filters. Kümmel himself says that in 1891 the number of the germs rose very rapidly, but fell again with equal rapidity ; as soon as only a few of the filters were cleaned, it fell at once to the normal number, and in three weeks all was over. But such periods of increased numbers of germs may be still shorter ; they sometimes last only a few days, and then, when the examination takes place only once a week, they easily escape observation.

Attention having already been drawn by the previous enteric-fever-epidemics to the possibility of the insufficiency of the filter-work during a cold winter, suspicion of course took the same direction when in January and February 1893 cholera-cases unconnected with one another occurred in all parts of the city. Introduction from Hamburg, which had played so great a part during the previous summer, was out of the question, for some of the patients had never left their homes, others had had no demonstrable intercourse with Hamburg, and, above all, the cholera had already completely died out in Hamburg. Of some of the last Hamburg cases it was even believed in Hamburg that the infection must have been caught in Altona or at least in the Hamburg suburb of Sankt Pauli, which had for some months past been supplied with Altona water. This fear even led to the closing of the outlets for Altona water in Sankt Pauli.

The question now was whether the said suspicion was really well founded, and the first thing to be done in order to answer it was to consult the record of the bacteriological investigations. Since the 1st of October 1892 Weisser had examined specimens taken from the pure-water-reservoir of the Altona waterwork almost daily. His figures at once showed that the number of the bacteria in the filtered water had increased for a very short time so early as the first week of December. On the 30th of December 1892 the number began again to rise, rose on the 12th of January 1893 to 1516, fell again for a while, and rose in the last week of January to from 1200 to 1400. After this there was no longer any doubt that some disturbances must have taken place in the working of the filters. When such disturbances occur, they of course never affect all the filters at once, but always only this and the other; it was necessary, therefore, in order to detect the fault, to examine the filters at once bacteriologically one by one. Unfortunately this could not be done, for the filters were not provided with apparatus enabling one to take the water just filtered from each of them. Only filter number 8 could be examined by itself; numbers 1 to 4, 5 to 7, and 9 and 10 poured their water into the pure-water-reservoir by means of common pipes, and could therefore be tested only in groups. According to Weisser's investigations the number of the germs went down to 1256 on the 30th and to 354 on the 31st of January.* From the 1st of February onward we have the results of the investigations of a bacteriologist who is appointed to conduct the daily investigations of the filters of the Altona waterwork. Since the 3rd of February it has been possible to examine filters 9 and 10, and since the 8th of February all the filters, separately. The numbers for February are given in the following table.

The figures of this table are of extreme interest. They show in the first place that filter number 8 and the pair 9 and 10 worked badly on the 1st of February. The separate examination of the latter two on the following day showed that number 10 was the one that filtered badly. The group, consisting of 5, 6, and 7, also failed to give sufficiently cleansed water. At the same time the number of germs per cubic centimetre in the pure-water-reservoir was only 154, a number which would scarcely have induced one to suspect a disturbance in the working of the filters; and yet at least two of them were working badly. But the bad result was so far concealed by the better working of the other filters that the collected water of the pure-water-reservoir showed only a trace of it, which could easily have been overlooked. During the previous days, on which much higher numbers of germs were

* The specimens to which these last numbers refer were not taken from the pure-water-reservoir but from a tap in the military hospital in Altona. The water of the pure-water-reservoir, being nearer the filters, would probably have yielded somewhat lower numbers.

found in the water in the city, the disturbances in the working of the filters must have been considerably greater. It is probable that, owing to the necessarily increasing deposits of mud on the filters as the process of filtration went on, they had abated down to the degree which was ascertained on the 1st of February.

THE GERM-CONTENT OF THE WATER OF THE ALTONA WATERWORK.

February.	Filters.										Number of germs in the water of the pure-water-reservoir per cubic cm.	Number of germs in the Elbe water before filtration per cubic cm.
	1	2	3	4	5	6	7	8	9	10		
1.		...				...		832		...	154	28,520
2.		88				212		550	908		142	35,340
3.		106				374		C.	76	636	110	40,920
4.		123				276		208	96	520	146	31,360
5.		176				206		544	84	362	105	33,480
6.		418				306		401	82	334	68	39,680
7.		234				204		446	94	C.	94	41,660
8.	50	22	40	24	28	136	146	368	84	152	130	25,560
9.	48	28	C.	32	54	194	152	182	64	122	72	44,140
10.	108	50	88	20	28	120	98	110	58	112	126	42,160
11.	68	60	76	78	36	140	C.	126	76	204	152	34,100
12.	72	58	240	60	38	110	288	80	70	282	82	26,040
13.	34	30	560	48	28	82	214	186	86	374	104	24,800
14.	40	46	354	24	18	52	164	142	46	364	142	34,080
15.	26	28	76	14	18	44	74	48	26	72	49	40,360
16.	38	26	84	24	22	52	76	60	C.	120	78	25,420
17.	20	36	156	C.	22	48	82	86	324	130	95	26,400
18.	26	18	102	54	32	54	112	72	82	126	91	26,440
19.	24	20	88	78	28	44	98	82	64	102	70	24,800
20.	26	22	70	104	24	36	96	88	34	78	46	19,840
21.	20	14	80	68	C.	34	96	44	30	152	50	34,720
22.	34	C.	46	62	158	34	68	36	64	...	42	18,250
23.	46	246	52	66	138	46	56	54	72	174	68	14,560
24.	22	42	32	36	72	22	72	76	34	44	54	11,080
25.	18	36	30	28	48	16	48	42	36	38	48	12,360
26.	14	20	24	21	34	12	40	36	28	34	32	9,370

(*N.B.*—C. means cleaning of the filter.)
The examination of the water of filter number 10 on the 22nd did not succeed.

Another point to be noted is the behaviour of the filters immediately after being cleaned. The table teaches that a filter that has been cleaned almost always works unsatisfactorily for some time after. From the numbers of the following months, which on the whole agree with those of February, and therefore need not be given here, it is evident that the renewing of the sand layer, which is necessary for every filter from time to time, causes a still greater disturbance than the simple cleaning. Filter number 10 for instance, the sand of which was filled up on the 13th of March, had 1880 germs on the 15th, and still 148 on the 24th. Filter number 8 had only 20 to 30 germs several days running before the renewal of its sand, 1364 immediately after, and 468 and 244 on the two following days.

For the rest, the daily investigation confirmed the previous experience that when a filter is working well it furnishes water with fewer than 100 germs per cubic centimetre.

These circumstances are of high importance for the future judging of the performances of a filter-work, and I shall have to return to them later on.

According to the investigations of the 1st of February, filter number 8 was the one that had been working worst, and the defect which had caused the disturbance of the filtration was therefore to be sought for in it. Fortunately it was discovered. On the 3rd of February the water standing on the sand layer of the filter was let off, and the mud layer removed. It was found, as Director Kümmel wrote me on the 4th, that the surface of the sand layer was frozen. According to Herr Kümmel's communications, the freezing of the sand layer took place as follows:—After a long period of frost somewhat milder weather had set in, and they wished to avail themselves of it for cleaning the filters. They began with number 8 at a temperature slightly below freezing-point, freed it of ice, and let off its water. Then, however, the temperature fell quite unexpectedly to 14 degrees below zero Réaumur (half a degree above zero Fahrenheit), and the surface of the filter was completely frozen, so that the cleaning could be finished only with the greatest difficulty. The filter was filled with water and left inactive for the usual time, after which, however, it hardly filtered at all, that is, hardly any water flowed through. Several days elapsed before the filter could be used at the rate of about 40 millimetres, after which it gradually improved, and could filter at last at about double that rate. This had happened from the 24th till the 26th of January.

The freezing of the surface of the sand layer during the cleaning of the filter, then, was found to be the root of the evil. The freezing up of the sand makes it utterly impervious to water, and therefore the filter did not work at all at first. Under the water and in contact with it the ice layer must gradually have melted at some places, and allowed filtration to proceed first at 40 and at last at 80 millimetres, that is, nearly the usual rate. This statement, however, means only that the filter gave as much water as it would have given with its full area at the rate of 40 and at the rate of 80 millimetres. But it did not filter with its whole area, probably only with a small part of it, for on the 3rd of February the surface of the sand was found almost completely frozen. From this it is to be inferred that the water must have passed through the pervious fraction of the filter at a rate proportionate to the loss of area. As it cannot be arithmetically determined how much the filtering area was diminished by the freezing of the surface of the sand, it is also impossible to state the rate of filtration with arithmetical accuracy. But, under the circumstances described, it will unquestionably have attained the maximum of which the difference of level between the unfiltered and the filtered water admitted, that is, the filter worked with the filtering area still at disposal as badly as possible.

It used to be disputed that the freezing of the surface of the sand could give rise to disturbances of the working, on the ground that a thin coating of ice that may possibly form under the water, the temperature of which is of course always above freezing-point, must soon melt. This case, however,—the first in which a timely and accurate investigation has taken place,—furnishes a palpable proof that the freezing of the surface of the sand does occur, and may disturb filtration in the most dangerous manner. In future therefore, we shall always have to take this possibility into account in the case of waterworks which have open filters.

But filtration may have the most formidable difficulties to contend with in winter without any freezing of the surface of the sand at all. In order to clean open filters in winter, the sometimes quite thick coating of ice which forms on the not yet filtered water standing on the sand must be removed. It is broken, and the ice is pulled piecemeal out of the water. The Altona filters have a comparatively very small area (somewhat over 800 square metres), and it is not excessively difficult to free them of their coating of ice, especially if, as is always done, the edge of the filter is constantly kept free of ice. The larger the areas of filters are, the more difficult is this incessant conflict with the ice that forms on their surface. The Berlin waterwork at the Stralau Gate has open filters which are four times as large as the Altona ones, and there the cleaning of the filters in winter seems to have almost insurmountable difficulties to contend with. From 1885 to 1891, during which years the water of these filters was bacteriologically examined in the Hygienic Institute in Berlin,* a period in which the water was insufficiently filtered occurred almost every winter. The reason was that it was impossible to clean the filters long after the mud layer had become too thick, because the water on the open filters was thickly covered with ice. The consequence was that the whole work had to be done by the few roofed filters, in which the water was not frozen, so that great irregularities in the working and maximum-rates of more than 200 millimetres an hour were often unavoidable.†

But the open filters of the Stralau waterwork are not the largest. The new Hamburg waterwork has open filters 7500

---

* Report on the Examination of the Berlin Water from the 1st of June 1885 till the 1st of April 1886, by Plagge and Proskauer, Periodical for Hygiene and Infectious Diseases, vol. ii. Proskauer "On the Quality of the Berlin Water during the period from April 1886 till March 1889," Periodical for Hygiene, etc., vol. ix. Proskauer "On the Quality of the Berlin Water during the period from April 1889 till October 1891," Periodical for Hygiene, etc., vol. xiv.

† Verbal Report of the Information given by the Manager of the Waterwork. Compare Periodical for Hygiene, etc., vol. ix., page 123; vol. xiv., pages 264 and the following.

square metres in area, that is, almost ten times the size of the Altona ones. The future must teach whether it is possible to free such large filters of ice and clean them so quickly that the surface of their sand does not freeze. In the light of previous experience this seems to me hardly practicable.

Winter, with its periods of frost, is not the only enemy of filtration. In summer also periods occur * in which it is hardly possible to keep the filters in good working-order. This happens when the water becomes rich in microscopic plants, and the consumption at the same time reaches its maximum. The vegetable micro-organisms have for the most part slimy integuments, which quickly stop up the pores of the filter, and can make it unable to perform its function in a few days. The increased consumption of water in summer does not allow the filters to rest long enough after being cleaned; they are worked too early, and at too rapid a rate, and the water they give is insufficiently cleansed in proportion.

Fortunately such disturbances, which occur frequently enough in almost all filter-works, do not always cause dangerous epidemics. The presence of infectious matter in the unfiltered water must happen to be simultaneous with the insufficiency of the filter-work, in order to bring about such a disaster. And then it is not the whole work that is defective, nor all the water that goes uncleansed through the filter; as a rule, only one or a few of the filters are at fault, and perform their function badly for a time. Even in the most unfavourable case, then, a filter-work allows only a part of the germs of infection to pass through. But that even these may suffice to cause epidemics is proved by the epidemics of enteric fever in Altona, the epidemic of enteric fever caused by the water of the Stralau waterwork in Berlin in 1889,† and the cholera-epidemic in Altona during the first two months of this year.

* Piefke's "Communications on Natural and Artificial Sand-Filtration," Berlin, 1881, page 9.

† "Experiments to Test the Efficiency of Sand-Filtration," by Fränkel and Piefke, Periodical for Hygiene, etc., vol. viii.

It has been already mentioned that the filtration of the Altona waterwork had been insufficient for a while so early as the first week of December. At that time Hamburg had been free of cholera for weeks, and the first isolated cases of the after-epidemic were just occurring; three cases occurred between the 5th and the 12th of December. There was therefore still a lack of infectious matter in the Elbe water, and for this reason the defective filtration was not followed by bad consequences for Altona. Meanwhile, however, the little after-epidemic in Hamburg developed (27 cases between the 19th and the 28th of December), infectious matter got into the Elbe, and when, towards the end of December, the disturbance of the Altona filter-works which I have described in detail set in, the germs of infection were spread over the city, and caused several scattered cases, of the introduction of which from elsewhere there was no evidence. Considering the small number of cholera-cases in Hamburg, the infecting of the Elbe water cannot have been more than slight; it seems too that only two of the filters were out of order, and to this circumstance it is to be ascribed that the outbreak of cholera in Altona was restricted to small dimensions, and soon ceased when the filter-work was put in order. Had Hamburg furnished more infectious matter, and had the disturbance in the working of the filters lasted longer, the epidemic would probably have assumed quite other dimensions. There would then have been reason to fear too that in Altona, the sewage of which also goes into the Elbe, a circulus vitiosus would develop, by means of which the cholera might survive the winter there, a result which had to be prevented by all possible means.

That infectious matter went from Hamburg into the Elbe is by no means a mere hypothesis, for, as has been mentioned in a previous communication,* cholera-bacteria were found in the water of that river. They were found in it not far below the mouth of the Hamburg main sewer. They were found

* "On the Present State of the Bacteriological Diagnosis of Cholera," Periodical for Hygiene, etc., vol. xiv.

also in the water of one of the two settling-tanks of the filter-work, that is, immediately before filtration. In the filtered water, indeed, they were not found, but they would probably have been found in it too, if much larger quantities of the water had been examined.

These are the facts I have to report concerning the relations between the filtration of water and cholera, with special reference to the Altona and Nietleben epidemics. The question now is what lessons we have to learn from them for the future.

The numerous and thorough investigations which have been made for some years past at the Berlin and Altona waterworks with regard to the process of filtration, and the bacteriological condition of the water before and after filtration, had led to the conviction that the filtration-rate of 100 millimetres an hour affords a sufficient guarantee for the satisfactory working of a filter-work. Further experience of those water-works, however, has shown that we do not gain very much by simply making this demand. For, with the now existing arrangements, most waterworks will not be able to fulfil it, and in point of fact they do not fulfil it. The principle must nevertheless be adhered to, that in future also a filtration-rate of 100 millimetres must be the first condition, but we must formulate our demand in more precise terms, and so far supplement it that the purpose aimed at may be attained with certainty. This is effected by the following somewhat extended demands :—

1. The filtration-rate of 100 millimetres must not be exceeded. In order to render this possible, each filter must be provided with a contrivance by means of which the motion of the water in the filter may be restricted to a certain rate, and by means of which also it may be ascertained at any moment whether this rate is observed or not.

2. Each filter-basin, so long as it is at work, must be bacteriologically examined once a day. It must therefore have an apparatus enabling one to take specimens of the water immediately after its leaving the filter.

3. Filtered water containing more than 100 germs capable of development per cubic centimetre must not be allowed to enter the pure-water-reservoir. The filter must therefore be so constructed that insufficiently cleansed water may be removed without mixing with the well filtered water.

To these sentences I have some further remarks to add.

Strictly speaking, the two last demands would alone suffice to avert the danger of the infecting of filtered water, so far as is possible with sand-filtration at all. But it seems to me questionable whether the demand of the daily bacteriological investigation of each filter is always to be insisted upon. If, owing to good construction and skilful treatment, a water-work regularly yields good results, the strict exercise of bacteriological testing could be restricted to times of danger, that is, to times of maximum-consumption of water, periods of frost, and when epidemics seem imminent; in the intervals a less elaborate testing, say the examination of the collected water every three days, would suffice.* For the times of less strict bacteriological testing, however, a guarantee for the regular working of the filters must be given by strict limitation and testing of the rate of filtration. As yet only a few waterworks possess apparatus for this purpose. The rate of filtration is generally calculated by ascertaining the proportion of the water filtered in 24 hours to the area of the filter. But everybody who knows the ordinary working of waterworks knows also that the demands made upon them in the course of 24 hours vary very considerably. At certain hours of the day very much water is consumed, at night, on the other hand, little. Unless the reservoir is large enough to balance these inequalities, this is effected by changes in the rate of filtration. The statement, then, that a waterwork filters at the rate of 100 millimetres, if based on the calculation indicated above, has but a very conditional value.

If the bacteriological testing of the filtration of water is to

---

* The weekly bacteriological investigation of the collected filtered water, now usual where the water is bacteriologically examined at all, is to be regarded as insufficient under all circumstances.

be carried out on a large scale, it must yield results in the shortest possible time. To this end it is advisable to keep the gelatine plates in the incubator at a temperature of 22 degrees centigrade, just as in cholera-investigations. The time of investigation may perhaps be still further considerably shortened by using agar plates at 37 degrees. In this connection I refer the reader to what I have written regarding the improved bacteriological diagnosis of cholera.

The opinion that filtered water containing more than 100 germs is not sufficiently cleansed has been completely justified by the experience of the Altona waterwork, which is confirmed by that of other works. Of course this statement is not to be understood to mean that water containing 101 or 105 germs is to be rejected without more ado. Each case must be intelligently judged of by itself, and the number 100 is merely intended to afford those called upon to form such judgments a basis founded on experience.

It is, strictly speaking, a matter of course that every filter must be provided with apparatus for the removal of insufficiently filtered water. Nor should I have expressly formulated this demand, if there were not still some filter-works which do not possess such apparatus. It is high time that this extremely dangerous defect be everywhere remedied.

Special prescriptions as to the cleaning and filling of the filters, the limit to which the sand layer may be allowed to waste, the removal of the first water after the putting on of new sand and after each cleaning, are not necessary, if the working of the filters is subjected to regular bacteriological testing, and the water which must according to the result of the bacteriological examination, be regarded as inadequately filtered removed. In this case the question whether in our climate open filters should always be replaced by covered ones may also remain undiscussed. It is the manager's affair to take care that the filtered water always fulfils the bacteriological demands. That construction and that treatment of the filters which yield the water freest of germs will always be the best. If, therefore, arrangements be discovered which render it possible to filter well with open filters in periods of frost, the open filters may continue to be used; if not, they absolutely must make way for covered ones. Each waterwork will have to construct its own rules with the help of bacteriological investigation; especially it will have to find out how long its unfiltered water requires to form a good filtering mud layer, how much water must remain unused after the cleaning, owing to its containing too many germs, how far the sand layer may be allowed to waste, etc. It is also the manager's business to ascertain the best remedy, if, as so often happens, too great demands are made on the waterwork, and regular filtration is thus rendered impracticable. In one case the only remedy will be the enlargement of the waterwork, in another the prevention of waste by the introduction of water-metres may suffice. All this, as I have said, may be left to the waterwork itself, if only it binds itself always to provide a bacteriologically sufficient water. In all cases, however, in which the bacteriological testing of a waterwork is declined, it will be absolutely necessary, in order to prevent mischief, to subject the work to the most vigilant supervision with regard to all the sources of defect alluded to above. But who is to undertake this supervision? Only the state can do it. And it not only can, it must, it is its duty. How many things does it not supervise already? Apothecaries' shops, hospitals, steam-boilers, factories with their contrivances for the protection of operatives, etc. all stand under state-supervision, in order to prevent injury to a few individuals by unskilfulness and negligence. But, if an accident happens to a waterwork, it is not a few individuals that are endangered, but the health and lives of thousands. Now that this conviction has inexorably forced itself upon us, we cannot possibly leave these things any longer to themselves, waiting till still more misfortune is wrought perhaps by cholera than in Hamburg and at Nietleben, or by enteric fever than in Altona and Berlin. It is high time to exchange the waiting attitude for that of energetic interference.

Our hitherto almost blind confidence in the filtration of water has been considerably shaken by the occurrences that have been discussed above. In future, therefore, when waterworks have to be constructed, it will be necessary to inquire whether it will not be better to choose other water instead of filtered surface-water. Almost the only substitute for surface-water that could be thought of in former times was spring-water; the use of underground water, as several unsuccessful attempts had taught, being unadvisable, owing to the iron it contained. But the question just raised has entered on a new stage of late, as it is now possible to remove the iron from underground water in a simple and inexpensive manner by airing and filtration. Thus a far superior rival has begun to contest the field with surface-water, and it seems that, thanks to this improvement, the old prejudices against underground water are disappearing more and more, for the number of places that have supplied themselves with such water has considerably increased of late. The first works of this kind were those of Halle, Leipsic, Dresden, and Charlottenburg. Their example has been followed by Norderney, Kiel, Bonn, Cologne, Elberfeld, Düsseldorf, Mannheim, Dortmund, Mülheim, Oberhausen, Barmen, Köthen, Krefeld, Linz, Pressburg, Pesth, and other towns.[*]

So far as the danger of infection is concerned, underground water affords absolute safety, and should therefore, if procurable in sufficient quantity, and not open to objection on account of chemical qualities, such as too great hardness or too large content of chlorides, always be preferred to surface-water.

I even think it desirable, and in certain cases necessary, that already existing works which filter river-water should be converted into works for the procuring of underground water. I should like to explain myself on this subject by reference to a definite example.

The Berlin waterwork outside of the Stralau Gate is in

[*] Salbach's "Report on the Experience made with Waterworks supplied with Underground Water during the last Twenty-Five Years," Dresden, 1893.

very unfavourable circumstances. At the time when it was built, the Spree, at the place where the water is taken, was exposed to comparatively harmless pollutions. In course of time, however, a material change has taken place in this respect. The city has spread far and wide up the river beyond the waterwork, so that the latter is now within its precincts. The water from a part of the sewage-fields in the north of Berlin flows into the Spree not very far above the waterwork; the river-traffic is much larger than it used to be; above the work lies Köpenick, with numerous washing-establishments. If one takes all this into account, one will hardly think the Spree water at the place where it flows into the waterwork better than the Elbe water below Hamburg. The construction of the work is faulty in several respects, in judging of which, indeed, it must not be forgotten that it is the oldest filter-work on the Continent. Of the difficulties it has to contend with in winter, owing to its open filters, I have already spoken by the way. Moreover, the demands made on it in the course of the year, nay even in the course of a day, are so unequal that its small reservoir cannot balance the great fluctuations, and that consequently a definite rate of filtration cannot possibly be maintained. If the work were not under the admirable management of the engineer Piefke, who has proved his ability equally in science and in practice, the enteric-fever-epidemic of 1889 would probably not have been the only one. The Waterworks-Department, in just recognition of the extremely dangerous situation of this waterwork, has taken care to have it replaced in time by another with covered filters on the Müggel Lake. This new work is to be started on the 1st of July next, and the Stralau work is to be stopped on the same day. Doubts have been raised, indeed, whether it will be possible to dispense with the Stralau work so soon. Should these doubts prove well founded, the case which I have here in my eye would happen. I do not believe that, if the danger of cholera should threaten us again at this time, anyone will have the courage to accept the responsibility for the continued working of so dangerously situated a waterwork. Then there would remain no other expedient than to take less dangerous water,—if possible, water quite free of danger,—instead of that of the Spree; and the only possible substitute would be the underground water. Fortunately the Stralau waterwork is as favourably situated for the use of such water as it is unfavourably situated for that of the river-water. Former investigations * of the ground in the neighbourhood of the waterwork showed that there is an extensive bed of gravel there, uncommonly rich in water, at a depth of about 15 metres. A well 1·5 metre in diameter, sunk into this layer of gravel by way of experiment, yielded 3000 cubic metres of water daily for a considerable length of time without the least abatement of its copiousness. This water was found to contain iron, but it was in great part out of the experiments with the water of this well that the method now used for ridding underground water of iron grew. It may therefore be relied on with certainty that it will be possible to free this water of iron. If the stream of underground water in that district were tapped by a sufficient number of wells, it would easily furnish as much water as the engines of the Stralau work could deal with. In order to rid the water of iron, it must be aired and then freed of the eliminated iron by filtration. The erection of an airing work would therefore be necessary. For filtration, on the other hand, no new work would be needed. This filtration is quite different from that which aims at freeing surface-water from micro-organisms. The flaky precipitate of ferric hydrate which separates itself from ferruginous water after airing is retained by pretty coarse sand or even by gravel, and at a very rapid rate of filtration. For this purpose the three covered filters of the waterwork would amply suffice; the open ones could be disused, and disturbances during winter would no longer have to be feared. I believe I have thus shown that this waterwork, which now furnishes water of doubtful value, could thus, with comparatively slight altera-

* Piefke's " Communications on Natural and Artificial Sand-Filtration," Berlin, 1881. " Soil-Filtration," Berlin, 1883.

tions, be transformed into one which would give equal quantities of faultless water.

Several other waterworks known to me are in similar circumstances, and could be improved on the same principles.

In order to discuss the relations of the filtration of water to cholera as exhaustively as possible, I must add a few words on artificial and natural filtration on a small scale, in contrast to filtration on a large scale, which I have been discussing hitherto.

For artificial filtration on a small scale, that is, for the wants of a family or a house, small filters of the most various construction are to be had. Not many of these rid water of micro-organisms, including infectious substances, and those which do,—for instance, those made of infusorial earth, argillaceous earth, asbestos, or cellulose,—keep germ-proof only for a few days, very soon deteriorate as regards quantity of yield, and require very careful treatment. No small filters are known to me which would suffice for practical purposes for a length of time, and I should not advise anyone to rely on small filters in cholera-times.

Natural filtration is a much better safeguard. When rain-water penetrates into the ground, and finally becomes underground water, it passes through much thicker layers, and at an infinitely slower rate, than river-water in artificial filtration through sand filters. If only the soil in question is fine-grained enough, soil-filtration is a very much more perfect process than artificial filtration. As a rule the soil consists of a much finer-grained material than the rather coarse-grained filter-sand, and it was therefore to be expected that underground water which had passed through sufficiently thick layers of fine-grained soil would be very poor in micro-organisms, or even free from them. This conjecture is confirmed by the investigations of C. Fränkel, who has proved that underground water, even in a soil the surface of which has long been strongly polluted, as is the case in Berlin, is absolutely free of germs. In other places also the same observation has been made. We have therefore no reason whatever to exclude the underground water, found almost everywhere, from use, even in inhabited places. On the contrary there is no water better filtered and consequently better protected against infection than underground water. All that is necessary is to make this most perfectly purified water so accessible for use in such a manner that it cannot be polluted and infected after purification. But it is inconceivable how grievously people everywhere still sin in this respect.

If one raises underground water by means of iron tube-wells, the possibility of any after contamination is excluded. The soil clasps the tube so tightly that no real disturbance of the filtering layers of earth is effected by the well. All fluids, even the most strongly polluted, must, before they penetrate to the depth from which the water is raised, pass through thick and effectively filtering layers, in which they are freed from infectious substances with absolute certainty. This may be especially relied upon when the well-tube is driven through an upper impermeable layer into deeper water-bearing strata of sand or gravel. Water got from such depths, it is true, will almost always be ferruginous. But this is just as little a reason for rejecting it in the case of single wells as in the case of large waterworks. All that is necessary is to free the water of iron on the principles already stated, that is, by airing and filtration; it will then be of first-rate quality, not inferior to the best spring-water. For this purpose the process described by Piefke,* which has been used with very good results in Hamburg in the case of a number of wells with strongly ferruginous water, is to be recommended.

Decidedly the most rational way of getting underground water is by means of iron tube-wells. It has, indeed, been

* Piefke's article "On the Utilisation of Ferruginous Underground Water for the Supplying of Towns" in Schilling's Journal for Gas-Lighting and Water-Supply, 1891.

objected that the supply they furnish gradually diminishes. Where this has been observed, the reason has almost always been that the metal sieve, intended to protect the tube against the intrusion of sand at its lower perforated end, was choked with mud or incrusted. It will be easy to remedy this, however, if the lower part of the tube is so constructed that the metal sieve can be changed when necessary.

Unfortunately metal tube-wells are still much too little used, and underground water is procured almost everywhere in the old but most irrational fashion by means of tank-wells. Such wells are so constructed that, if the bottom of the well is deep enough, and lies in soil that filters well, only well filtered water gets into the well from below, that is, from the underground water. But from above the well is almost always exposed to the most dangerous pollutions. Very often such wells are quite open, or but scantily covered; but, even if they are closed above by masonry, or even iron plates, chinks and rents always form in the uppermost layers of soil surrounding the well, which dry up in summer and are exposed to frost in winter, and these chinks and rents allow unfiltered water to enter from above. Masonry and even cement-work are loosened and torn by the same influences, and do not keep off the water which flows to the well from the surface of the soil. But just this water may be most foully polluted. The well-water itself is used at the well for washing linen, cleaning chamber-pots, and similar purposes. Thus contaminated and possibly infected, it flows by the shortest way it can find through chinks and rents in cover and walls back into the well-tank. Besides, wells generally lie deeper than the surrounding ground. Consequently the filth of overheaped dunghills, gutters, etc., flows to them; the rain too washes the dirt deposited round human dwellings, often from a considerable distance, into the wells. This is evidently the reason why epidemics emanating from wells are observed most frequently after heavy showers of rain. An outbreak of cholera in Altona, in the environs of a well into which filth had got from above, furnishes a proof that cholera-infection also may emanate from a tank-well contaminated in this manner. I shall have to report on this case in another place. See p. 93.

Tank-wells, no matter how constructed, must no longer be tolerated in future, if they are exposed to pollutions of the kind I have described, or even if there is reason to suspect the possibility of such contamination.

It will not be easy, indeed, to effect the simple abandonment of already existing tank-wells, however badly constructed and dangerously situated. But that is not always necessary. It is generally possible to alter a tank-well by comparatively simple means, so as to exclude all danger of contamination from above. It is only necessary to give it, in the shape of effectively filtering layers of soil, the same protection against contaminating influxes as the simple tube-well possesses, or at least a protection approximately equal to that. To this end one can fill the well-tank up to the highest water-level with gravel, and lay fine-grained sand over it up to the brim of the well. It is of course presupposed that the well has an iron pump-tube, or, if not, is provided with one before being filled up. By these means a tank-well can be converted into a tube-well, and it has even one advantage over an ordinary tube-well, namely, that its lower end dips into a layer that offers hardly any resistance to the underground water. If it is necessary to preserve the tank of the well with its store of water in order for instance always to have a certain quantity of water at disposal for fire-quenching purposes, a structure of masonry or of iron girders, able to bear the protecting covering of sand, must be laid over the highest water-level. The sand-covering, however, must not be less than two metres thick. It is very advisable too not to erect the pump immediately over the well, but at a suitable distance from it, and to connect it with the well-tank by means of a lead pipe. This prevents the water of the well which is polluted beside the pump by rinsing and washing from oozing into the ground too near the well.

Wells protected in this or in a similar way by effectively

filtering layers afford the same security against the infecting of water as the sand-filtration of the large waterworks, nay, strictly speaking, a still greater security, as they are not exposed to the numerous above-mentioned disturbances, and especially not endangered by frost.

People are now everywhere endeavouring to perfect the supplying of water on a large scale to the highest possible degree, but they should direct their attention to the procuring of water on a small scale also, and seek to limit the spread of cholera to a minimum, so far as it depends on water, by improving wells in the manner I have indicated. Just in this respect a great deal still remains to be done.

The Cholera in Germany during the Winter of 1892-93.

In the second half of October 1892 the great cholera-epidemic which had raged in Hamburg since the 16th of August seemed ended; and, as all the centres of infection, of which Hamburg was the source, were also extinct, one was at liberty to hope that, so far as Germany was concerned, the danger of cholera was over for the time. This hope, however, proved deceptive, for an after-epidemic broke out in Hamburg soon after, which restricted itself, indeed, to very small dimensions, but again gave occasion to several transmissions of the pestilence, which caused more or less violent outbreaks of cholera in Altona and in the asylum at Nietleben near Halle. At the instance of the Prussian Ministry of Religious, Educational, and Medical Affairs, the Institute for Infectious Diseases has under my guidance made these two last-mentioned epidemics the subject of the most thorough investigation possible, which has led to several not unimportant results, and of which I shall therefore give a minute description here.

The outbreak of cholera in Altona is so closely connected with the after-epidemic in Hamburg, and the latter presents some features which bear so directly on what I have to say, that I must say a few words about this after-epidemic first. We may expect a minute description of it in the official report which Professor Gaffky of Giessen is preparing at the instance of the imperial authorities.

1. THE AFTER-EPIDEMIC IN HAMBURG.

The first epidemic had begun on the 16th of August 1892, and might be regarded as ended on the 23rd of October. Isolated cases occurred, however, so late as the 9th and 11th of November.* The number of cases during this epidemic was 18,000, that of the deaths 8200. The beginning of the after-epidemic may be dated the 6th of December. The last case belonging to it was ascertained on the 4th of March, the second-last on the 11th of February. This second epidemic numbered only 64 cases and 18 deaths. If we leave the two isolated stragglers of the first and the last case of the second epidemic out of account, we find that both had an almost equal duration of somewhat more than two months. But what a tremendous difference in the intensity of these two epidemics, despite the agreement as regards their duration!

One might suppose that the time of the year had an influence, and that the summer epidemic was favoured by the high temperature, and therefore attained so large dimensions, whereas the winter epidemic was hampered in its development by the prevailing cold. But this explanation falls to the ground, if we consider that in the coldest period of the same winter the low temperature was unable to prevent an outbreak of cholera at Nietleben in which the proportion of the sick to the healthy was much higher than in the summer epidemic in Hamburg. The history of earlier cholera-epidemics also shows that the most violent outbreaks occasionally occur in winter. In this case therefore one cannot regard difference of temperature as an essential factor. Just as little can it be maintained that Hamburg had been thoroughly attacked by the first epidemic, and, in virtue of a kind of immunisation, no longer offered material for more than a slight after-epidemic. Though I agree with explanations of this kind in general, I should like nevertheless to point out, as regards the case in question, that on several

* Reincke, "The Cholera in Hamburg," German Medical Weekly, 1893, Nos. 3 and 4.

former occasions Hamburg suffered from considerable successive epidemics within a single year, and that this is not to be wondered at, because its population fluctuates much. Add to this that a large number of the inhabitants had left the city at the beginning of the epidemic, and that, on their return in October, these fugitives brought the city a not inconsiderable quantity of unattacked material, without any consequent increase of the epidemic.

We must therefore cast about for some other explanation of that striking phenomenon, and I believe that such an explanation may be given as follows :—

I have already pointed out on a former occasion that cholera shows two quite different types in its outbreaks. One of these is marked by the explosive course of the outbreak. The graphic representation of such an outbreak gives a curve with its first leg rising steep and high, and its second falling with almost equal steepness. The graphic representation of the second type, on the other hand, presents the appearance of a curve rising only a little above the base line. Hamburg shows these two types in its last epidemics in a quite extreme form. The curve of the summer epidemic has the appearance of a very high and sharp triangle with a quite short base, that of the after-epidemic rises so little above the base that it almost coincides with it.

The first type is produced by the sudden and equable strewing of the infectious matter over a place. The epidemic which then breaks out takes an explosive course, and its graphic representation forms a curve high and steep in proportion to the quantity of the infectious matter which has been, so to speak, sown out. This type of epidemic, however, is subject to the conditions that the local distribution of the cases be somewhat equable, and that no immediate connection between them can be recognised. One must not, indeed, imagine the distribution to be quite equable and planlike, even when this type appears in its purest form; for the sowing is hardly ever quite equable, and the soil on which the seed falls is not equally well fitted in all its parts to develop the germ. Individual predisposition, cleanliness, alimentation, density of population, habits of life, etc. exercise an influence not to be underrated. An equable sowing, such as is presupposed in this type, can be effected only by means of something that can act on all, or at least most of the inhabitants of a place simultaneously, such as air, water, soil, and means of nourishment. But neither air, soil, nor means of nourishment have hitherto been proved to be the media of explosive outbreaks of cholera. Insects too, which have with good reason been thought of, are out of the question here, for cholera-explosions are not very rare in the cold season, when transmission by insects is quite impossible. Small groups of cases may, indeed, be caused by infected articles of nourishment, and it is not to be denied that insects may play a part in such cases by carrying the infectious matter to such articles; but the sudden infection of whole communities, such as we see so often in the case of cholera, cannot be explained in this manner. Only water then remains, and the fact that water can indeed be the bearer of the germ of cholera, not only for single groups in the population of a place, but for whole communities and even large cities, has been proved by former epidemics, and once more with quite special conclusiveness by the outbreaks of cholera in Hamburg, Altona, and Nietleben. But just this opinion that the infectious matter is carried by water has been objected to on the ground that the distribution of the disease in such epidemics was too unequal; the infected water, say the objectors, goes into all dwellings, and yet one finds houses and even whole streets in the districts supplied with such water which were little or not at all attacked by cholera; whereas, if the water had been the cause, all the persons who came in contact with it must have been attacked according to a certain percentage. This objection would be quite correct, if the virus of cholera were a quite equably distributed substance dissolved in the water, if all the persons in question had drunk exactly equal quantities of it, and if the susceptibility to the virus were equally great in all persons. But we know perfectly well that not one of these conditions was fulfilled. There is not the shadow of a doubt, and this has always been emphasised by bacteriologists especially, that individual predisposition for cholera varies greatly. It seems also hardly necessary to point out that the possibility of infection by water must vary as much as the relations of different persons to water vary. One person drinks no water at all, he comes into contact with it only indirectly through its use in housekeeping, and he is therefore proportionally less exposed to the danger of infection than another who drinks water. But, as regards the latter too, it is not indifferent whether he drinks much or little water, at what time he drinks it, whether with a full or an empty stomach, whether the functions of his stomach and intestines are in order at the time or not, whether excesses have been committed, etc. The distribution of the infectious matter, that is, the cholera-bacteria, in water is also to all appearance not such as many suppose. The most recent bacteriological investigations* show that the cholera-bacteria exist in considerable quantity in infected water perhaps only exceptionally, and it is therefore by no means necessary that every drop or every mouthful of infected water contain cholera-bacteria. It is also very doubtful whether they are from the first quite equably distributed in water, and whether, if they are, they remain so. It is easy to believe that, like other bacteria, they occasionally get attached to solid objects, the inside of a water-pipe for instance, especially when the motion of the water is temporarily or permanently slackened. They may then perish at the place to which they have attached themselves; but, under more favourable circumstances, they may also multiply, or be swept away by stronger currents. In general the unequal motion of the water in a system of water-conduits must exercise a considerable influence on the forwarding of the cholera-bacteria, and may alone bring it about that in one range of pipes many, in another few cholera-bacteria are washed into the attached houses. If, further, the houses supplied by the latter range

* The Periodical for Hygiene and Infectious Diseases, vol. xiv., page 336.

happen to be inhabited by well-to-do people, who, in virtue of their habits of life, offer few points of attack to cholera, it may happen that whole rows of houses, nay even streets, are spared by the disease, without our being entitled to regard the fact as a proof that the water is not infected.

The same question which is occupying us here has been discussed already in former times, and it may not be superfluous to remind the reader of what Farr[*] very justly answered to those who disputed that the cholera which broke out in the district supplied by the East London waterworks in 1866 was caused by the infecting of the water. It was not yet possible at that time to demonstrate the insufficiency of the purification of the water in the filters by bacteriological investigation, but there was a circumstance which proved that the filtration must have been temporarily defective. In the water that flowed from the pipes in a number of houses eels had been found, which furnished an unambiguous proof that unfiltered water had got into the pipes.[†] Farr concluded that the virus of cholera had got into the water-conduits and through them into the houses by the same way as the fish from the river Lea, which was strongly contaminated with fæces. His opponents answered that there had been no cases of cholera in many of the houses which got their water from the East London waterwork, and that therefore the water could not have played a part in spreading the pestilence. Farr's answer to this objection ran as follows:—

"Eels, as we have seen, were found in the water of a certain number of houses in East London. To argue that in hundreds of other houses no eels were found, and that therefore the company never distributed eels in the district, would be absurd. The fallacy of such reasoning is transparent. It assumes the form:—if no eels are found in the waters of a certain number of houses, none exist in the

[*] "Report on the Cholera-Epidemic of 1866 in England," London, 1868, page xxx.

[†] The examination of the workmen and engineers of the waterwork showed that unfiltered water had really been repeatedly pumped into the conduits. See Farr's Report, page xxii and the following pages.

waters of any houses. As the eels are limited in number, they cannot be distributed universally; and the fact that they were discovered in one house and not another would depend on laws and circumstances so intricate as to make the ascertained distribution anomalous, but not necessarily more anomalous than the distribution of the lower forms of organised matter, to which the phenomena of cholera in man are due."

The second type of cholera is distinguished from the first not only by the form of the curve but also by some other characteristic qualities. The distribution of the cases is not equable; well marked foci form, in which the disease nestles. Nor do many cases suddenly occur at such a focus; on the contrary they follow one another, form chains as it were; and a direct connection can very often be detected between the several cases of the focus. One person for instance, who has come from some other place, falls ill; a few days later a member of the family in which the patient was nursed; then in rapid succession, but often also at considerable intervals, other members of the family, inhabitants of the same house, neighbours, persons who visit the house, etc. The infection may be carried from the first focus, and new foci formed in other parts of the city, or in neighbouring places, where new chains of successive cases form larger or smaller groups.

Here too one must not demand that each individual link in the chain of cases be distinctly recognisable. It is impossible to lay bare the intercourse of human beings in all its finest ramifications, and to find out every person who has come into direct or indirect contact with a cholera-patient. If cholera-cases were from the first of so severe a nature that they must all come to the knowledge of medical men, the power of the cholera-patient to infect another would cease as soon as his attack was over; and, if the infection were effected only by immediate contact, it would indeed be possible, notwithstanding the complex relations of human intercourse, to ascertain the individual links of the chains, with but few exceptions, with the aid of bacteriological diagnosis.

But we know now that some cases of cholera are so mild that they generally escape recognition; we know also that the attack of cholera, strictly so called, is only the most striking part of the illness, and that the infectious matter may be contained in the evacuations of the patient both before and after it, that is, at a time when his intercourse with others is not yet or no longer regarded as dangerous. Finally, it must also be taken into account that the transmission by no means always emanates directly from the patient, but is much more frequently effected indirectly by means of linen, clothes, bed-clothes, articles of nourishment, insects, etc. If one considers all this, one will certainly find it explicable that in a thinly sown rural population, with communications of little complexity, the connection between the cases is pretty completely discovered, but that in large towns and cities it is only now and then possible to detect the mutual relations of the links in so manifoldly intertwined and often ramified a chain. Another circumstance which renders it specially difficult to get a general view of this kind of cholera-propagation is that it restricts itself almost exclusively to the lowest densely crowded and constantly fluctuating strata of the population, attacking the better situated classes only here and there. And yet this type of cholera is pretty easily recognised by the focal grouping of the cases at single spots. On careful inquiry one always finds in such cases cholera-nests, in which the bringing of the disease from another place, and the further spreading round step by step are distinctly evident.

It would be a mistake, however, to suppose that cholera must always assume one or other of these two types, for it is obvious that both may be, nay often must be, combined. It is especially frequent that the first type, which generally appears uncombined at first, combines in the course of its development with the second type, and finally loses itself entirely in it. It also happens that a local epidemic begins with the second type, retains it till the infectious matter accidentally finds its way into the water, and then causes little circumscribed explosions, varying according to the nature of the water-supply, or suddenly infects a whole district or even, under certain circumstances, the whole place.

It is also necessary to mention that the form of the cholera-curve alone is not decisive as to the type. Even in the case of a water-epidemic the curve may remain very low, if the sowing of the cholera-bacteria by the water is very sparse. On the other hand it is not impossible that numerous and almost simultaneously originated foci may give the curve a form approaching more or less nearly to that of the first type, so that the second type may assume the external form of the first. The fact is that, in judging of cholera-epidemics, one must not, if one wishes to avoid errors, fall into the mistake of schematizing, but must examine each local epidemic by itself, in order to be able to decide how much of it belongs to the one type, how much to the other. The present epidemic has given us uncommonly instructive examples in this respect.

The Hamburg summer epidemic, for instance, belonged exclusively to the first type at first. From the very beginning the cases were without connection, and pointed at first to the harbour as the sole source of infection. Owing to the connection of the Hamburg water-supply with the Elbe, and indirectly with the harbour, a general explosion was even then to be feared, and unhappily this fear was not disappointed. Towards its end the epidemic passed into the second type.

The Hamburg winter epidemic, on the other hand, maintained the form of the second type almost pure during its whole duration. The tendency to form foci characterised it from the first.

One of these foci had its centre in the Neustadt (New Town), a second in the district of Sankt Georg (St George), and the third in the suburb of Sankt Pauli (St Paul's). Whether all three foci were connected could not be ascertained, but it is not probable that this was the case, and that the disease was carried, for instance, from the first focus in the Neustadt to Sankt Georg and Sankt Pauli. It rather seems likely that the two first foci originated in undetected stragglers of the summer epidemic. The summer epidemic was ended, as already stated, on the 23rd of October. But cases of genuine cholera were ascertained on the 9th and 11th of November, and they are not likely to have been the only ones. If, then, the after-epidemic began on the 6th of December, the interval between the two Hamburg epidemics was not more than four weeks at most, and, considering that, it is unnecessary to assume that the first case or cases of the after-epidemic came from elsewhere. Nor do I know where they could have come from, for the cholera was everywhere extinct at that time.

Whether the Sankt Pauli cases may be regarded as a focus is open to doubt. It is extremely probable that some of them had their origin in Altona; others were possibly connected with the focus in the Neustadt, so that but very few remain unaccounted for.

It is a very characteristic feature of the after-epidemic that the patients belonged without exception to the lowest strata of the population. They were people without work and without homes; drunkards living in beggars' inns and spirit-shops; itinerant hawkers who sold matches, sausages, or the like, and whose trade led them too into such haunts; sailors, dock-labourers, persons under police-arrest, etc. In all the cases except eight relations to persons who had suffered from cholera, and from whom the infection might have been directly or indirectly caught, were proved. This proof, indeed, was due only to the extremely thorough investigation of each case by the sanitary police. A superficial inquiry, such as used to be made under similar circumstances, would certainly not have detected the connection, and a new apparent cholera-riddle would have been added to the many of former times. A common cause, such as the influence of the soil, the water, or the like, certainly did not exist in this epidemic. The water-supply was out of the question, for the district affected did not coincide, as in summer, with the sphere of the water-supply. The soil might have seemed open to suspicion, as the disease was restricted to certain localities. In this case too, however, the place could not be the decisive factor, but, on the contrary, the people resident on the place, for the disease always ceased the moment the patients and the suspected were removed. Had the infecting agent been inherent in the locality, fresh cases must have occurred, despite the removal of the infected persons, among those who freely visited the houses in question. The only remaining explanation is that the disease was transmitted from person to person, and this explanation is decidedly supported by the chain-like connection of most of the cases. It is always necessary, however, to bear in mind that cholera-infection differs entirely from that of small-pox, measles, etc., in which simple contact or even a short stay in the sick-room suffices to let infection take effect. Such immediate transmission happens only occasionally, and is to be assumed, I believe, only when several cases of cholera occur in succession in one family at intervals corresponding to the period of incubation. Something of this kind happened in the Hamburg after-epidemic, in the course of which two families had four cases each. In all the other cases the infection seems to have been indirect, and all attempts to discover by what roundabout way the infectious matter got from one person to another failed. This entirely agrees with what has been observed in outbreaks of cholera on board ships for the transport of emigrants, pilgrims, or troops, in which, among people densely crowded together, and living under bad sanitary conditions, the disease dragged along in a loosely connected series of cases for weeks. One of the most characteristic instances of this kind is the cholera-epidemic on board the Italian emigrants' ship "Matteo Bruzzo,"* an instance so instructive that it ought not to fall into oblivion.

Though water did not play the part of a general factor in the after-epidemic, its powerful influence as a propagator of

* Conference for the Discussion of the Cholera Question [second year], German Medical Weekly, 1885, No. 37A, page 21.

cholera was not entirely absent, for there is no doubt that it had to do with the outbreak of cholera among the crews of two ships that lay in Hamburg harbour.

The first of these two ships was the Spanish steamer "Murciano," which lay at first at the Asia Quay, near a closet which is said to have been used by a Hamburg workman who was ill of cholera. On the 8th of January two of her crew had to be taken to a hospital as cholera-patients; the rest of the crew were then required to leave her, and, on more careful investigation, four other cholera-cases were found among them. She was then taken to the Strandhafen (Beach-Harbour), where she was disinfected, and her frozen closets were thawed. There she lay beside the steamer "Gretchen Bohlen," among the crew of which, consisting of negroes, cholera broke out on the 15th of January, just three days after the "Murciano" had been laid beside her. From this ship also two persons were taken to the hospital severely ill, and four mild cholera-cases were detected only on further examination.

When the first cases occurred on board the "Murciano," the first thought was that of immediate infection through the use of the above-mentioned closet. This assumption, however, was contradicted by the circumstance that six of the twenty-four persons of whom the crew of the ship consisted, and of whom it was not even certain that they had used the closet in question, fell ill at once; whereas not a single case of cholera occurred among the numerous dock-labourers going up and down on shore, for whom also the closet lay convenient. It was much more probable that the infection had been caught not directly by using the closet, but indirectly by the flowing of its contents into the water of the harbour, which was frequently used for drinking and cleaning in the ship. The quays of the Hamburg harbour have sewers, which are not connected with those of the city, but open into the harbour each at the end of its quay. All the liquid filth of these sewers, including the contents of the water-closets connected with them, flows into the Elbe, and is washed up and down by the ebb and flow of the tide beside the ships lying at the quays. In this manner the contents of the water-closet in question, including cholera-fæces perhaps, may have got into the ship by a pretty short way through the medium of the water.

Here we have to do with the very same circumstances which in all probability caused the cholera-epidemic in Hamburg harbour last summer. At that time it was the barrack of the Russian emigrants at the America Quay from which quite inadequately disinfected, or rather undisinfected, fæces, and water polluted by the washing of dirty linen, flowed through the sewer of the quay into the harbour. These tributes were by no means unimportant, for several hundreds of emigrants were arriving daily, who had to stay for several days in the barrack before they could be sent on. At the time when cholera broke out there consequently were, on an average, a thousand emigrants in the barrack, many of whom took advantage of the opportunity of cleaning their stores of dirty linen and clothes afforded them by the interruption of their journey. The assumption that the Russian emigrants brought the cholera to Hamburg has been objected to on the ground that no cholera-cases occurred among them before the outbreak in Hamburg harbour. It is true that severe, clinically unmistakable cases of cholera were not observed among the emigrants, but does that prove that they brought no infectious matter of cholera with them? A large number of them came from districts which were suffering severely from cholera; and who can maintain, in view of that fact, that there were not mild cases among them, or convalescents who had cholera-bacteria in their fæces for two or three weeks after recovery, or that there were no cholera-fæces among the bedclothes, linen, etc. which they carried with them in large quantities? It would have been surprising if, under such circumstances, no infectious matter of cholera had been brought by such emigrants, and if, after it had once found its way into the emigrants' barrack, from the barrack into the sewer, and from the sewer into the harbour, the harbour-population had not been infected. Hamburg harbour, with its arrangements as they then were, formed an uncommonly vulnerable point of attack for the threatening cholera-invasion, and cholera could not but gain a footing there, if any unfortunate accident offered it an opportunity. The attempt to prove that the infection was brought from some other source,—from French ports for instance,—has failed, and there is no alternative but to lay the blame on the emigrant-traffic, which, as has been shown, offered overabundant opportunity.

As regards the Spanish steamer "Murciano," there was a doubt, at least at first, whether the infection was due to the water, but there was from the first no doubt on that head in the case of the second ship, the "Gretchen Bohlen." She had arrived in Hamburg harbour on the 5th of January; the "Murciano" was laid near her, disinfected, and cleaned on the 12th, and cholera broke out on board the "Gretchen Bohlen" on the 15th. The crew, consisting of seventeen negroes, had been free of cholera till then, had had no other opportunity of catching the infection, but had, as has been ascertained with certainty in this case, drunk water directly from the Elbe in large quantities. As the cholera took exactly the same course on board the second ship as it had taken on board the first, the assumption that the outbreak on board the latter had really been caused by infected water was rendered still surer.

One of the most striking peculiarities of the Hamburg after-epidemic is its low mortality. It amounted to 28 per cent.; whereas the usual mortality of cholera, as is well known, is about 50 per cent. I believe, however, that this exception from the rule was probably only apparent. The older cholera-statistics counted only cases which showed well-marked clinical symptoms, that is, severe cases. Milder cases of diarrhœa with vomiting and cases of simple diarrhœa were called cholerine, and generally left aside. The Hamburg after-epidemic was the first in which bacteriological diagnosis was carried out as completely as possible, and every case in which cholera-bacteria were found registered as cholera. These cases include not only such as were formerly regarded as suspiciously like cholera and called cholerine but also such as showed quite unimportant clinical symptoms, nay, none at all, and were investigated only because the persons in question had been in contact with indubitable cholera-patients. This epidemic, in short, was the first in which not only the clinically but also the etiologically suspicious cases were examined, which procedure revealed the extremely important fact that the latter too include a certain number of cases of cholera-infection which can be detected as such only with the aid of bacteriological investigation. During the great summer epidemic, and even towards its close, when the cases became rarer and rarer, bacteriological investigation was restricted to clinically suspicious cases, partly because there was a lack of time for such investigations and of men able to conduct them, partly also because there was no real occasion for them. In the winter epidemic, however, it very soon appeared that this did not suffice. Even if the clinically suspicious were at once rendered harmless by isolation, sporadic after-cases kept constantly cropping up, which proved beyond doubt that the infectious matter had not yet been completely got rid of. At my advice the evacuations and the bacteriological investigations connected with them were more and more widely extended. In giving this advice I was guided by the idea that, as the surgeon cuts into the healthy flesh when he wishes to remove a malignant growth with certainty, so also the extirpation of the cholera-germ must be effected in the healthy flesh, as it were, if it is to have a chance of success. Consequently not only single individuals suspected of cholera, but all who could be reasonably supposed to have caught the infection in the same way as they or from them, were taken to the evacuation-station and bacteriologically examined there, whether their fæces were diarrhœic or not. Under this method it appeared that there really were among the apparently healthy individuals persons whose fæces were hardly diarrhœic, nay quite normal, and nevertheless contained cholera-bacteria. That such persons must also be regarded as infected by cholera, and consequently as bearers of the infectious matter of cholera, is obvious, for among the very numerous other investigations of the fæces of healthy persons and of persons suffering from the most various diseases which have in course of time been made in bacteriological laboratories, especially during the last epidemics, nothing of this kind has ever been found; such cases happened exclusively among people among whom even clinically indubitable cases of genuine cholera had occurred, and of whom one must therefore assume that they had had an opportunity of catching the infection. How it comes that the same opportunity of infection can effect so various gradations of the disease, and whether this is due to the differences of individual predisposition alone, or also to other influences hitherto unknown to us, must remain undecided for the present. It is to be hoped that further observations and investigations will give us the answer to this question too. At any rate it is now certain that among a number of persons who have been exposed to cholera-infection, the resultant cases may show the whole scale from the severest and rapidly fatal cases down to the mildest imaginable, demonstrable only by bacteriological investigation. I regard this experience as one of the most important additions to our knowledge of Asiatic cholera, both from the practical and from the theoretical point of view.

It is practically important for the following reasons :—

If one rests content, as formerly, with rendering only the clinically suspicious and afterwards bacteriologically ascertained cholera-cases harmless by isolation and disinfection, one will doubtless succeed in some cases in extinguishing a cholera-focus in course of development; but in other cases, especially in densely crowded city-populations, and under circumstances so unfavourable as those in Hamburg were, the efforts to destroy all the cholera-germs would be vain. And this method is open to the special objection that precisely the mildest cases, which escape investigation, are the most dangerous of all as regards transmission. This can be most simply explained by some examples. On board each of the two cholera-ships in Hamburg harbour two people fell ill with clinical symptoms which could not but lay them open to the suspicion of cholera; of course they were at once isolated. If now, after disinfecting the ships, one had left the rest of the crews, who seemed to be quite well, unmolested, eight persons whose fæces contained cholera-bacteria would have had an opportunity of carrying the infectious matter once more from place to place in the environs of Hamburg harbour. Suppose that the crews had not been foreigners but natives, had gone after passing muster to their respective homes, had given rise perhaps at first to the development of mild cases there which escaped recognition, while they themselves had never been ill of cholera from the clinical point of view, the cholera might have been carried in this manner to a considerable distance, and subsequent investigations need not have detected the slightest evidence of its origin. The following circumstances, which occurred in a beggars' inn in Hamburg, seem to me specially noteworthy in this respect. A diarrhœa-patient from this inn, which had had eight cholera-cases during the great epidemic, four of them fatal, was bacteriologically examined on suspicion of cholera on the 26th of December, and found to be really suffering from that disease. In consequence of this discovery as many of the inmates of the inn as could be laid hold of were also examined, and among them was found a man who had no clinical symptoms justifying suspicion of cholera, but who had cholera-bacteria in his fæces. When I visited the same inn some days later, it was full of people, by far the greater part of whom had no occupation, some of whom, as they said, had ceased to expect to find occupation, and were on the point of leaving Hamburg again. They were from the most various parts of Germany; natives of the Prussian province of Saxony and of the Kingdom of Saxony were specially numerous among them. One can easily imagine how cholera can be carried by such people from one place to another, and how its traces may not rarely escape the most careful investigation. Had not the cholera been so energetically pursued in Hamburg into its most secret hiding-places, and every trace of infectious matter that could be found rendered innocuous, I am convinced that it would have proved impossible to get the upper hand of the kindling matter that had been spread so copiously over the city. In this manner, however, the single sparks were quenched before they could kindle new conflagrations. Every spark of course could not be discovered at once, and the several chains ran on in secret; but they became rarer and rarer, and had also to cease to glow at last. I believe the case which was ascertained on the 27th of May, after an interval of four months, to have been an offshoot of such a chain. It completely justifies the continued caution observed in Hamburg, and thoroughness in the bacteriological investigation of suspicious cases of illness even after the apparent close of an epidemic.

In the theoretic province the demonstration of the mildest cholera-cases can be turned to account in two directions.

In the first place, we now know that a not inconsiderable number, nay, if we take the case of the two ships as typical, the majority of persons infected by cholera, show so insignificant symptoms of disease that, under ordinary circumstances, that is, without bacteriological investigation, they would certainly be regarded as in good health. This disposes at once of all the difficulties which have hitherto been found in the fact that human intercourse can propagate cholera, even when only healthy persons are in question. It really not rarely happens that no notoriously sick people, or inanimate articles, such as linen etc., laden with infectious matter, have come, or at least can be proved to have come, to the infected place. Such cases have been interpreted as indicating that, if cholera can be carried without cholera-patients or their fæces, a cholera-patient is no more fitted to propagate pestilence than any other fraction of human intercourse, and such interpreters have then quite consistently gone the length of declaring cholera-patients and their fæces comparatively harmless. How overhasty this interpretation of the transmission of cholera by apparently healthy persons is, is now obvious. Whoever maintains that cholera has come to a place without cholera-infected persons or their fæces having anything to do with it, must now prove that there were no cholera-cases of the mildest type among the apparently healthy people who came to the place, and that no articles soiled with cholera-fæces were brought to it.

The experience gained in the Hamburg after-epidemic also gives us the key to the results of the attempts hitherto made to infect human beings with cholera, and of unintentional experiments of the same kind. The first observation of this nature was made by Macnamara.* He reports that of nineteen persons who accidentally drank water polluted with cholera-fæces five fell ill of cholera. Unintentional infections have since occurred in the laboratory of the Imperial Office of Health † and in that of the city-hospital in Dantzic. To these must be added the well-known intentional experiments in Munich, Vienna, and Paris. Macnamara does not tell us whether there were severe and fatal cases among those observed by him. But the other cases of infection mentioned above, which could only be effected by pure cultivations of cholera-bacteria, all have the peculiarity that, when the cholera-bacteria appeared in the fæces of the infected persons, more or less severe diarrhœa set in, though, except in the one case in Paris, the real severe symptoms of cholera did not develop. That is, those symptoms which we must ascribe to the resorption of the virus produced by the cholera-bacteria were wanting. One might infer from this that the cholera-bacteria are, indeed, able of themselves to produce more or less severe diarrhœa, but that they cannot cause cholera dangerous to life. This inference, however, is not correct, for, if we take the repeatedly mentioned outbreaks of cholera on board the two ships in Hamburg harbour into comparison,

* German Medical Weekly, 1885, No. 37A.
† See the same Article.

and Macnamara's observation into account, we find that of a certain number of persons, who are simultaneously exposed to infection, only a definite fraction take the disease in a severe and another in a mild form, while the rest remain well. On board the Hamburg ships, which afford the surest ground to go upon in this connection, two out of twenty-four persons took the disease in a severe and four in a mild form in the one, and likewise two out of seventeen severely and four slightly in the other. If, then, in the case of the isolated laboratory-infections, and of the intentional infections affecting only a few persons, only mild cases of illness occurred, this agrees entirely with what the experience hitherto gained would lead one to expect. Even if those experiments had had an absolutely negative result, they would not, in the least, disprove the specific nature of the cholera-bacteria, for, among the groups of persons infected in the usual way, the majority do not fall ill either. If such experiments are to gain the intended purpose, they must be adapted to the natural circumstances in every respect. A considerable number of persons must expose themselves to infection with cholera-bacteria. Some of them must swallow the bacteria with empty stomachs, and simultaneously with a large quantity of cold water. Others must, after diarrhœa has set in, and cholera-bacteria have appeared in their fæces, commit errors of diet, and eat articles of food known by experience to favour the outbreak of cholera, and so on. If, in an experiment arranged in this manner, and with fresh cultivations of full virulence, only mild cases resulted, then, but not till then, one would have to examine further under what special conditions the severe cholera-symptoms occur, and whether special aids, independent of the qualities of the cholera-bacteria, and of the variations in the condition of the digestive organs, are needed to produce such symptoms. Till then there is no reason to doubt the present opinion that the cholera-bacteria are able of themselves to produce cholera-symptoms, mild in one case and severe in another, according to the individual predisposition of the infected. This of course by no means robs the experiments hitherto made of their importance; they afford a most valuable contribution to our power of judging of the capabilities of the cholera-bacteria, though they do not prove what those who subjected themselves to them intended to prove by them.

From the tremendous focus which had developed in Hamburg during the months of August and September 1892, the pestilence was carried to nearly 300 places in Germany and abroad. But the winter epidemic too, strikingly minor as it was both in extent and intensity, was not without its offshoots.

In the beginning of January a man travelled from Hamburg to Schwerin, and died there of cholera.

Four cases were carried from Hamburg to Elmshorn, which place had also received six cases from Hamburg the summer before.

A case, which, indeed, has not been completely cleared up, is said to have occurred near Hamburg, at Osdorf in the sub-district of Pinneberg.

At Neuhof, in the island of Wilhelmsburg in the Elbe, a ship-carpenter died of cholera on the 6th of January. He had worked in that part of Hamburg harbour in which the above-mentioned steamer, "Murciano," which had cholera on board, lay. This case is therefore probably referable to the same source of infection.

Three cases of cholera occurred at Schulau, and one at the immediately adjacent town of Wedel. These cases could not be traced directly to Hamburg, but indirect relations existed in the water of the Elbe. Wedel and Schulau lie on the right bank of the Elbe, about twenty kilometres below Hamburg, and eight below the Altona waterwork. They get their water from wells, but a sugar-factory at Schulau takes its water from the Elbe by means of pipes of its own. The place from which the water is taken is close to the landing-bridge where the Elbe barges which bring beetroots, coals, etc. to the factory touch. At the same place the river makes a little curve, and has a pretty strong current. Old corks and other floating objects, which are washed ashore there in considerable numbers, show that, despite the long distance, the contaminations from the Hamburg and Altona sewers reach the place still in a pretty concentrated condition. Before flowing into the factory, the water is filtered, but only through a layer of gravel, which hardly keeps back even the coarser filth. This water was drunk by the factory-hands; one of them fell ill, and died on the 10th of January, and another fell ill on the 26th. The former lived at Schulau, the latter at Wedel. Soon after the death of the former, two of his children were attacked by cholera. The linen and clothes of this workman, whose illness was taken at first for a case of poisoning by meat, were sent to relations in Silesia. Fortunately, however, parts of the corpse had been sent at the same time to Kiel for bacteriological examination, and, when cholera was diagnosed there, there was still time to order the destruction of the dangerous objects by telegraph, before they had done further mischief.

None of these transmissions from Hamburg caused cholera-outbreaks of any considerable intensity. Real epidemics developed only in Altona and at Nietleben, and with these we shall now have to occupy ourselves more minutely.

II. THE WINTER EPIDEMIC IN ALTONA.

The cholera-epidemic in Altona simultaneous with the great Hamburg epidemic had ended, like the latter, towards the end of October. Sporadic stragglers on the 4th and 29th of November* seem to have been connected with it. Then followed a period of at least apparent freedom from cholera, which lasted till the end of December. At this time the after-epidemic in Hamburg had reached its culminating point with five cases on one day (the 26th of December), and now the pestilence began to reappear in Altona too. This time, however, it bore a character quite different from that of the previous epidemic. At that time by far the

* Wallichs' "The Cholera in Altona," in the German Medical Weekly, 1893, No. 10.

most of the patients were connected with Hamburg, and had in all probability caught the infection on Hamburg soil, or from persons who had caught it there. Now, on the contrary, the patients were with a few exceptions persons whose illness could not possibly be explained in that way. They were, for instance, workmen in fairly good circumstances, whose work did not require them to go to Hamburg; women of the middle class; little children who, owing to the cold weather, had not even been out of the house; a one-year's-service-soldier; inmates of the hospital who had been confined to bed for weeks, and had not come into contact with anybody outside; and a prisoner who fell ill after twelve days' solitary confinement. There was no connection whatever between any of these cases. The simultaneous epidemic in Hamburg was restricted, without exception, to the lowest strata of the population, and had evidently nestled itself in at certain spots; whereas in Altona only one of the patients, a homeless woman, belonged to the lowest class; all the rest were persons who could not have caught the disease by immediate transmission from uncleanly people crowded together in unhealthy rooms. Evidently the cholera in Altona was of a quite other type than in Hamburg, and the only explanation of the phenomena was that the infectious matter was scattered all over the city, but only scantily. The first thought could not but be of the water-supply, for, if accidents had happened to that, they could easily have betrayed themselves by producing the phenomena in question. Cholera-germs, washed out of Hamburg through the sewers into the Elbe, would have been scattered with the insufficiently filtered Elbe water all over Altona, and could not but have produced disconnected cholera-cases. That this suspicion was well founded, and that disturbances in the working of the filters at the waterwork had actually taken place immediately before the outbreak of the epidemic; further, of what nature these disturbances were, and the fact that cholera-germs, that is, cholera-bacilli, were found in the water of the Elbe just below Hamburg and in the water of the Altona waterwork before filtration; all this is discussed in detail in a previous treatise* to which I refer the reader.

Small epidemics are in some respects more advantageous subjects of etiological investigation than great ones. The individual cases can be much more thoroughly investigated, each for itself, and their mutual relations can be detected in a manner which is absolutely impracticable in great epidemics. The articulation of a small epidemic therefore remains, as a rule, transparent and intelligible in its connection. The Altona epidemic with which we are now concerned also possesses these advantages. It was to be expected from the first that, if the infectious matter got scattered over a whole city, and a number of persons fell ill in consequence, the epidemic would not remain restricted to these primary cases. Wherever circumstances favourable to the direct or indirect transmission of cholera exist, the primary cases will be followed by secondary ones. But in a great epidemic such secondarily developing continuations of the infection can be separated from the whole, and their manner of originating correctly recognised, only in exceptional cases; generally they lose themselves unrecognised in the mass. In the small Altona epidemic, however, it proved possible adequately to distinguish even the secondary infections, which, as it were, sprouted out from the basis of the water-epidemic.

In three places groups of cases originated within families.

In one family the father suffered first from diarrhœa; two children died a few days afterwards of cholera; and two days later the mother, who had been occupied with the nursing of the children, died also.

In another family a child of seven months fell ill first, and died on the 1st of February; then, two days later, a foster-child of a year and a half and a girl of thirteen fell ill; and

* "Water-Filtration and Cholera," in the Periodical for Hygiene and Infectious Diseases, vol. xiv.

after a three days' interval a girl of six and the father. The last three recovered.

The third family-epidemic happened in the family of a baker. Two women fell ill on the 2nd of February. As they must have caught the infection at the same time, it is probable that their illness was preceded by a mild case of cholera which escaped recognition, and formed the starting-point of this group of cases. On the 3rd and 5th of February, a female relative and a daughter of the baker fell ill. All the cases of this group were mild.

In the hospital also a small group of cases seems to have formed by direct transmission. The first cholera-patient there was a woman who had been treated for pulmonary phthisis in the chief pavilion for four weeks. The second was a typhus-convalescent in the third barrack, and the three following cases were in the department for mental diseases. These three last cases may probably, especially after the experience gained in other establishments for the treatment of the insane, be regarded as connected with one another.

The most interesting secondary group of cases developed in the part of the city called Ottensen. There is a group of houses there (See diagram 1),* bounded by four streets (Rothestrasse, Papenstrasse, Grosse, and Kleine Brunnenstrasse), which consists of two parts. All the western houses, except the corner-houses of the Papenstrasse and Grosse Brunnenstrasse, are connected with the Altona water-conduits, and are marked on the plan with a W. The eastern part consists of several long courts, accessible from the adjacent streets, with small houses on both sides. They were erected after the great Hamburg fire of 1842, in order to provide shelter for the burned-out people as soon as possible.† Since that time they have been inhabited by poor tenants. These

---

* I owe the sketch for this plan, the information about the borings, about the underground water in Altona, and other valuable communications, to City-Architect Stahl of Altona.

† Wallichs' "The Cholera in Altona," in the German Medical Weekly, 1893, No. 10.

courts, the population of which numbers about 270 souls, bear the name of "lange Jammer" ("long Misery"). The adjacent streets are provided with sewers; and clay pipes, furnished with gullies, go from the sewers to the courts, so that the removal of the sewage is tolerably provided for.

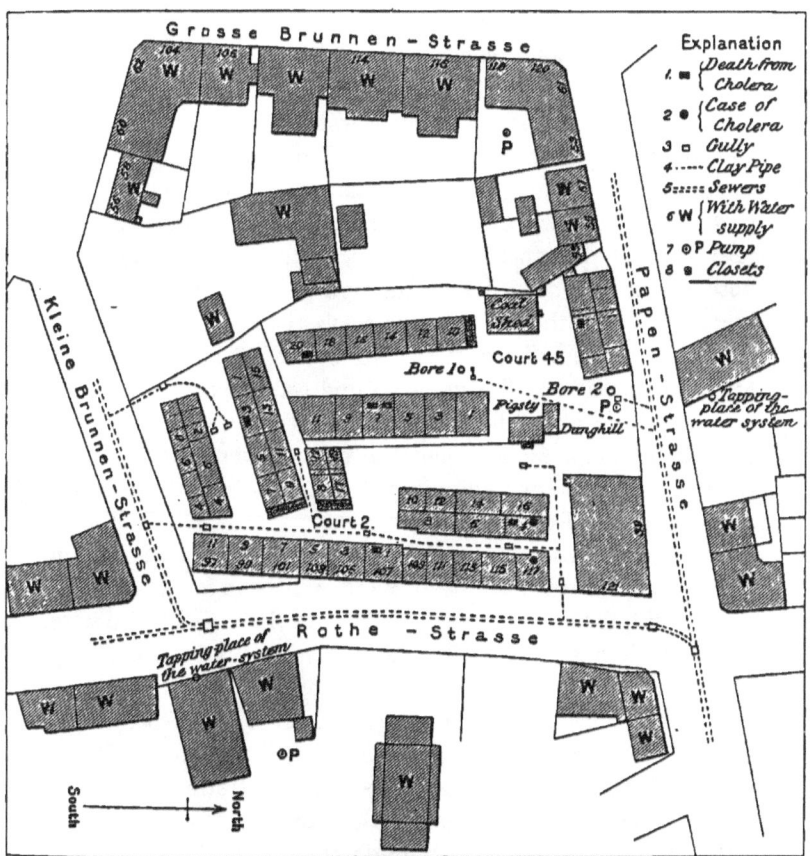

FIG. 1.—Plan Explanatory of the Well-Epidemic in Altona.

The houses of the "lange Jammer" are not connected with the water-conduits. The reason of this is that the Altona water-system belongs to a private company, which sells the water. The company's water was too dear for the poor inhabitants of the "lange Jammer," who had to get their water elsewhere; and the only source from which they could procure it was an old pump-well beside the entrance to court number 45 from the Papenstrasse (See diagram 1). To the situation and construction of this well, which was so fatal to the inhabitants of the "lange Jammer," I shall return later on. As the "lange Jammer" has no water-pipes, it has to do without water-closets too.

The removal of the fæces is effected by means of closets provided with barrels. They seem to be sufficiently numerous, and are arranged in groups at five different places. In order to ascertain something about the nature of the ground in this part of the city, two bores have been made in court 45,—one near the well, the other almost thirty metres further south. At both places the bore passes through a layer of humus and humus mixed with sand a metre thick, then through clay and clayey sand one and a half to two metres thick, then through fine-grained sand to the underground water, which was found at a depth of about five metres. This agrees entirely with the observations made when the sewers were laid in that district. The soil accordingly did not differ from that of the adjacent parts of the city.

As regular observations regarding the fluctuations of the underground water have been made in Altona for some years past, I will, for completeness' sake, state their result also here, so far as it has to do with the cholera-focus in the "lange Jammer." The level of the underground water is measured in bores which lie in a line perpendicular to the Elbe, that is, running from north to south. Of these bores number V. is nearest the cholera-focus, and also almost at the same level. It may therefore be assumed that the fluctuations of the underground water in that bore do not essentially differ from those at the place in which we are interested here. On the whole the fluctuations of the underground water in the soil on which Altona stands are very slight, which is obviously due partly to the fact that the city lies pretty high above the Elbe, so that its underground water is not exposed to swells caused by the fluctuations in the level of that river, partly to the circumstance that the laying of the sewers sank the underground water to a definite level, and prevents its rising above it. At bore number V., the brim of which lies 27·45 metres above normal zero, the following measurements of the underground water-level were taken. The numbers indicate the distance of the underground water from the brim of the bore in metres, the increase of the number showing the sinking, its decrease the rising of the water.

From these figures it appears that the level of the underground water fluctuated between 9·38 and 9·83 in 1891, and between 9·46 and 10·02 in 1892. This gives a difference of 45 centimetres in 1891 and of 56 in 1892. In both years the level was highest in spring and lowest in autumn and winter, which is the case almost all over North Germany. In 1892 the level was on the whole lower than in 1891, owing in all probability to the fact that 1892 was a dry year.* About the end of the year the underground water had reached its lowest level. According to the well-known theory one might regard this as a circumstance favourable to cholera. This is inadmissible, however, because, if for no other reason, the cholera died out again in Altona before this factor ceased to act, for the table shows that the low level of the underground water lasted into February. Besides, it would be hard to see why just one group of houses in Altona should be so strikingly influenced by the underground water, while many other houses of quite similar situation and construction showed no trace of any such influence.

Even during the summer epidemic cholera-cases occurred in the "lange Jammer." The first happened on the 29th of August. The patient was a worker in tobacco, of whom it was stated that he had not left Ottensen, and that he had no relations whatever with Hamburg. His was therefore one of the comparatively few cases which formed the Altona

* According to van Bebber (see Annals of Hydrography and Maritime Meteorology, January, 1893), the quantity of rain etc. that fell in Hamburg in 1892 was 190 millimetres below the average.

The Cholera in Germany during the Winter of 1892-93.

1891.		1892.		1893.	
Jan. 17	9·60	Jan. 2	9·75	Jan. 7	9·97
,, 24	9·55	,, 9	9·70	,, 14	10·07
,, 31	9·46	,, 16	9·74	,, 21	10·05
		,, 23	9·71	,, 28	10·11
		,, 30	9·65		
Feb. 14	9·38	Feb. 6	9·73	Feb. 7	10·02
,, 21	9·41	,, 13	9·59	,, 14	10·06
,, 28	9·45	,, 20	9·56		
		,, 27	9·58		
Mar. 7	9·53	Mar. 5	9·55		
,, 14	9·59	,, 12	9·49		
,, 21	9·60	,, 19	9·50		
		,, 26	9·46		
Apr. 4	9·58	Apr. 2	9·57		
,, 11	9·56	,, 9	9·60		
		,, 16	9·62		
		,, 23	9·53		
		,, 30	9·58		
May 2	9·40	May 7	9·59		
,, 9	9·47	,, 14	9·58		
,, 16	9·43	,, 21	9·59		
,, 23	9·36	,, 28	9·57		
,, 30	9·45				
June 6	9·43	June 4	9·64		
,, 13	9·50	,, 11	9·64		
,, 20	9·48	,, 18	9·73		
,, 27	9·44	,, 25	9·56		
July 4	9·38	July 2	9·72		
,, 11	9·50	,, 9	9·73		
,, 18	9·53	,, 16	9·75		
,, 25	9·53	,, 23	9·76		
		,, 30	9·75		
Aug. 1	9·66	Aug. 6	9·77		
,, 8	9·67	,, 13	9·78		
,, 15	9·64	,, 20	9·80		
,, 22	9·66	,, 27	9·80		
,, 29	9·58				
Sept. 5	9·62	Sept. 3	9·81		
,, 12	9·60	,, 10	9·83		
,, 19	9·60	,, 17	9·85		
,, 26	9·54	,, 24	9·85		
Oct. 3	9·56	Oct. 1	9·87		
,, 11	9·62	,, 8	9·90		
,, 17	9·64	,, 15	9·90		
,, 24	9·60	,, 22	9·91		
,, 31	9·67	,, 29	9·93		
Nov. 7	9·83	Nov. 5	9·94		
,, 14	9·64	,, 12	9·94		
,, 21	9·70	,, 19	9·95		
,, 28	9·80	,, 26	9·95		
Dec. 5	9·70	Dec. 3	9·98		
,, 12	9·82	,, 10	10·01		
,, 17	9·86	,, 17	9·96		
,, 24	9·67	,, 24	10·02		
		,, 31	9·98		

epidemic proper. The case remained isolated. A second case which occurred on the 4th of September had no connection with it. The patient was a huckster or pedlar, who had been carrying on his trade in Hamburg during the preceding days, and had undoubtedly caught the infection there. This case, too, had no further consequences, and one would have thought that, as the cholera had been brought to the "lange Jammer" twice, and yet had not spread there, the local conditions were not very favourable to cholera. All the more surprising was it that, shortly after the beginning of the after-epidemic, nine cholera-cases with seven deaths occurred there in one week. Two houses had two cases each. The other cases all remained isolated, and were distributed pretty equably over the houses of the "lange Jammer."

When cholera-cases became so strikingly frequent in so narrowly limited a district of the city, inquiries were at once set on foot as to the causes to which the origination of the cholera-focus might be due. The surprising fact, already alluded to above, that in a city otherwise so excellently supplied with water there was a group of houses at the place in question with no water-supply, and that the cholera-focus was strictly limited to these houses, came to light. The neighbourhood of the "lange Jammer" remained absolutely free of cholera for a considerable distance round. The same thing which had been experienced on a many hundred times larger scale in Hamburg the summer before, namely, that the spread of the cholera coincided to a hair's-breadth with the sphere of a water-supply, had thus repeated itself here in miniature. The source of the water-supply in this case was a well, which of course became the centre of the inquiry.

The well is in court 45, at the spot indicated above (See diagram 1). It occupies, as such wells generally do, the lowest part of the court. Its elevation is 28 metres above normal zero; that of bore number 2, which is nearest to it, $28\cdot11$; that of the house-corner a little further west of it $28\cdot26$; that of bore number one $28\cdot58$; that of the closets near the latter $28\cdot46$ to $28\cdot54$; that of the house-corner east of bore number one (marked 1 in the plan) 28·61 ; that of the pigsty and the dunghill 28·52.

This being the situation of the well, all fluid matter on the surrounding surface naturally had to flow towards it. Under ordinary circumstances this was provided against by the clay pipe leading to the sewer, which clay pipe is the continuation of a gully near the closets, and goes from there in a straight line pretty close past the well to the street. At the well itself there is also a gully, which, with a short conduit of its own, carries the water that flows past during pumping likewise to the sewer.

The wall of the well is of bricks. The well is closed at the top by wooden planks, covered with a thin layer of earth and with stones. The depth of the water in the well is somewhat more than a metre ; its level corresponds to that of the adjacent bore number 2.

From this description it will be seen that the well possesses the same qualities as most such wells, and that, if it had been examined before the epidemic, there would have been no occasion to designate it as bad. And this actually happened. In May 1892 366 wells were examined in Altona, and 92 of them found unfit for use. In the list of wells before me, however, the well beside the Papenstrasse, which afterwards became so dangerous, is not among the condemned ones. This example shows with signal clearness how little confidence the examinations of wells usual nowadays deserve, and that a tank-well which has been used perhaps for years without disadvantage may under special circumstances suddenly become a source of infection. The construction of tank-wells in fact is such, as I have explained in a previous treatise (Water-Filtration and Cholera), that the entrance of infected fluids from above can never be prevented with certainty. And something of this kind must have happened to the well at the Papenstrasse. The infectious matter cannot have got into the well-water from below, that is in the underground water; for the bottom of the well lies in fine-grained sand of good filtering quality, which, moreover, is protected against contaminations from the surface of the ground by a pretty thick impermeable layer of clay. Besides, if the underground water had been the bearer of the infectious matter, not only this but also other wells, at least the neighbouring ones, must have been similarly surrounded by cholera-cases, whereas they were not. The corner-houses, for instance, Papenstrasse 59–61 and Grosse Brunnenstrasse 118–120, are not supplied with water, but use a well, and yet not a single case of cholera occurred in them.

The only remaining possibility therefore is that the infectious matter entered the well from above, and unmistakable indications that this was really the case were found on closer inquiry. For, after the wooden covering had been removed, wet stripes of dirt appeared on the side of the well where the sink and the gully are, and these stripes went down from the top to the water, and visibly marked the way which the dirty fluid from above had taken. Gaps and fissures must have formed in the masonry at the place where the stripes of dirt began, and in the thin layer of earth and stones outside, so that liquid matter could flow unhindered from the surface of the ground into the well. As a rule this fluid will have been only water which had been taken from the well itself, and had flowed along beside it during pumping, or when rinsing and washing were going on at its side. But even then infectious matter could have got into the well-water, if, for instance, linen or vessels soiled with the fæces of enteric-fever or cholera-patients had been cleansed at the well. That this did not happen sooner, especially during the summer epidemic, was due solely to good luck. As matters stood there, a cholera-focus might have developed suddenly in summer. It is possible too that the winter outbreak was caused in this manner, but it seems to me more probable that cholera-fæces or water used in washing cholera-linen had been poured out in court 45 near the well, and got into the well in the manner I am about to explain. In summer, it may be assumed, such a fluid would have flowed through the gullies into the clay pipe and through it into the sewers, or the dry ground would have absorbed it, and not allowed it to flow far. But in winter, when the ground was frozen almost a metre deep, and the gullies were frozen too, so that they could not receive any more fluid at all, the only way left for dirty water poured out in the court was that prescribed it by the natural fall of the ground, and this was the way to the well, which is situated at the spot where the ground is lowest. That this was really the case is also indicated by the fact that, when the court was cleaned and sprinkled with carbolic acid immediately after the occurrence of the first cholera-cases, the water of the well, into which no disinfectants could have got directly, is said to have smelt for a while of carbolic acid. It is further noteworthy that no dirt-stripes were visible on the inner wall of the eastern side of the well, where the clay pipe passes it, so that no contaminations can have flowed into it from that side. Besides, the clay pipe, which, it is true, lies two metres deep at the place, and consequently below the influence of frost, must have been empty at the time, for it received nothing from the frozen gullies.

As the weather played so essential a part in this case, it will not be unimportant to gain some exact knowledge of the meteorological conditions of the period under consideration. I give the data in question according to the observations of the Marine Observatory, which Professor van Bebber has kindly placed at my disposal.

This table shows that a short period of thaw which lasted from the 15th till the 21st of December 1882 was followed by a period of persistent cold. It began on the 21st of December, and lasted till the 24th of January. During this period the gullies must have been frozen and unable to perform their function. The outbreak of cholera in the " lange Jammer" began on the 21st of January, and ended on the 1st of February. It is therefore likely that the infecting of the well took place some days before, that is, within the period of frost, when dirty water could not ooze into the ground or flow through the gullies, but had to find its way

The Cholera in Germany during the Winter of 1892–93. 97

above ground to the well. On most of the days, indeed, the cold was so intense that such fluids too must have been very soon converted into ice, and could not flow far; but there were days, the 14th of January for example, on which the temperature was only very little below freezing-point, and on such a day the contamination of the well may have taken place.

At any rate, whether it happened in the way I have tried to explain here, or in some other similar way, cholera-

	December 1892.			January 1893.			February 1893.		
	Temperature °C.		Rain, &c.	Temperature °C.		Rain, &c.	Temperature °C.		Rain, &c.
	Min.	Max.	Mm.	Min.	Max.	Mm.	Min.	Max.	Mm.
1	0·4	4·6	5·9	− 8·1	− 5·1	3·7	3·9	4·5	1·0
2	− 2·5	1·9	0·9	− 7·9	− 5·3	0·0	1·1	4·8	4·5
3	− 6·4	0·8	12·1	− 13·1	− 7·3	0·0	− 3·1	2·9	0·0
4	− 2·1	3·1	3·4	− 14·8	− 9·5	6·4	− 4·6	− 1·3	0·0
5	− 1·9	3·6	4·5	− 4·9	− 1·8	0·0	− 8·9	− 2·4	0·0
6	− 4·9	0·5	0·3	− 4·1	− ·3·4	0·0	− 4·2	− 1·4	0·0
7	− 3·1	− 0·9	0·4	− 11·8	− 3·4	0·0	− 5·4	0·9	2·6
8	− 3·1	0·3	0·2	− 13·9	− 9·4	0·0	0·6	2·5	5·0
9	− 0·3	1·6	0·0	− 9·2	− 7·4	0·0	0·1	3·6	4·6
10	− 4·8	0·6	0·0	− 7·9	− 3·4	0·0	1·9	3·6	2·6
11	− 10·5	− 4·4	0·9	− 9·1	− 1·9	0·1	2·6	4·6	2·8
12	− 5·4	0·6	4·9	− 14·9	− 6·4	0·2	− 1·3	6·3	10·2
13	0·6	1·9	0·1	− 9·8	− 5·7	6·2	− 4·9	1·1	19·5
14	− 1·9	1·9	4·7	− 5·9	− 0·4	0·2	0·0	2·6	17·3
15	0·5	3·6	2·9	− 11·9	− 4·3	3·3	3·1	7·0	0·1
16	3·0	5·1	0·4	− 15·1	− 7·9	5·0			
17	2·8	5·8	1·9	− 13·7	− 5·6	0·5			
18	5·2	6·6	0·0	− 18·4	− 9·4	0·2			
19	4·8	7·8	0·4	− 18·2	− 14·6	0·0			
20	3·9	8·1	0·3	− 18·2	− 8·6	3·3			
21	− 1·0	5·1	0·3	− 3·7	0·8	0·0			
22	1·1	3·1	0·0	− 8·8	− 0·8	0·1			
23	− 4·2	2·4	0·0	− 6·6	− 4·8	0·1			
24	− 7·7	− 1·3	0·0	− 10·7	− 1·9	4·8			
25	− 8·7	− 5·1	0·0	1·7	2·6	0·0			
26	− 7·7	− 2·6	0·0	0·8	3·6	0·0			
27	− 3·4	− 0·2	1·0	− 0·6	1·1	0·1			
28	− 2·0	− 0·1	1·2	0·7	1·5	0·0			
29	− 0·3	0·6	0·2	− 3·8	1·8	0·8			
30	− 1·1	0·8	0·1	1·1	2·9	0·7			
31	− 6·8	0·0	1·1	1·9	3·9	3·2			

G

bacteria must have got into the well; for they were actually proved to be in the well-water. The case in question is one of the rare ones in which the circumstances were of such a nature that the investigation could be made early enough. Had the well-water been examined some weeks after the outbreak of the disease, as usually happens in such cases, nothing suspicious would have been found. As, however, in this case suspicion pointed to the well from the first, the investigation took place so early as the 31st of January. It was conducted in the manner described in a previous treatise;* and, as cholera-bacteria were found in very large quantity in the very first specimens mixed with peptone and kept at incubation-temperature, the original water must have been pretty rich in them too.

One litre of the water taken from the well on the 31st of January was kept in a place with a pretty equable temperature of three to five degrees centigrade, and was examined from time to time with a view to ascertaining whether it contained cholera-bacteria. The investigation yielded an affirmative result on the 2nd, 3rd, and 17th of February. The cholera-bacteria had therefore remained alive in well-water for eighteen days under the conditions stated. Specimens of water afterwards taken from the well itself and examined contained no cholera-bacteria.

The well was closed on the 26th of January, and one case of cholera occurred on the 27th, one on the 28th, one on the 29th of January, and one on the 1st of February,—all cases lying within the incubation-period (the last patient had been suffering from diarrhœa for some days), and consequently to be attributed to infection by the well-water.

How the cholera found its way into the "lange Jammer" could not, in spite of all efforts, be ascertained. I therefore conjecture that, among the isolated cases, scattered all over the city, that had occurred since the end of December, one or another had happened in the "lange Jammer," but had

* "The Bacteriological Diagnosis of Cholera," in the Periodical for Hygiene and Infectious Diseases, vol. xiv.

remained unrecognised, owing to the unremarkable character of its symptoms. The inhabitants of the "lange Jammer" do not belong to the class of people who seek medical aid even in a case of diarrhœa, and the assumption that a mild case of cholera had run its course among them unnoticed is therefore quite justifiable. If fæces of such a patient got into the well, the little epidemic needed no more complicated explanation. But, if anyone refused to admit such an assumption, one could ascribe the two first cases of the 21st of January to the general Altona epidemic, and the infecting of the well to one of them. If this explanation is correct, the well-epidemic began only on the 24th of January, and consisted of seven cases with six deaths.

There was another noteworthy phenomenon in the Altona epidemic. It lasted from the 23rd of December 1892 till the 12th of February 1893, and comprised 47 cases with 27 deaths. The mortality, therefore, amounted to 57·4 per cent. This figure is rather high, but not extraordinarily so. It is an extremely striking fact, however, that an epidemic prevailed almost simultaneously in the adjacent Hamburg, which, with 64 cases and 18 deaths, had a mortality of only 28 per cent., that is, not quite half so high. The difference must appear even greater than these percentages express, if we consider that the Hamburg patients were exclusively ill-nourished, low sunk people, in great part addicted to drinking, whereas the Altona ones were of all classes of the population. Why was the cholera in Altona so very much more malignant?

One's first thought might naturally be that the cholera-investigations were conducted in a much more thorough style in Hamburg than in Altona, and that the cases ascertainable only by bacteriological investigation, though they existed in Altona as well as in Hamburg, were not found out in Altona. Now it must be admitted that this circumstance may have had a certain influence, especially in the beginning of the Altona epidemic, for only the clinically suspicious cases were bacteriologically investigated at first, and some mild cases may have escaped detection. Afterwards, however, investigations were repeatedly made *en masse*, in the prison and among the troops of the garrison for instance, without clinically unsuspicious cases of cholera with bacteriological characteristics only, such as were detected in Hamburg, coming to light. The Altona epidemic, therefore, really differed from the Hamburg one in quality. This singular fact becomes almost more striking, if one takes the separate groups of cases into account. In one family, for instance, three of four patients died, in another one of four, in a third not one of five, and in the "lange Jammer" seven of nine. These differences must have been due either to external factors, such as soil, state and situation of dwellings, etc., or to individual conditions, or to the nature of the infection, or, finally, to the behaviour of the infectious matter, that is of the cholera-bacteria. The two first possibilities are excluded in this case, for differences of soil have not been discovered, and there is equally little reason to attribute the phenomenon in question to other external factors, for, as regards dwellings, cleanliness, and nutrition, the Hamburg victims were in much worse case than even the inhabitants of the "lange Jammer." Just as little reason is there to suppose that the members of the family in which three out of four cholera-patients died, or the cholera-patients of the "lange Jammer," were personally in worse condition than the homeless people and the drunkards in Hamburg. The influence which the nature of the infection and the quality of the infectious matter may have exercised therefore seems to be the factor most worthy of consideration. It is possible that cholera assumes a specially dangerous form when transmitted by water, as it was in Altona, and especially in the "lange Jammer." The malignant character of the great summer epidemic in Hamburg, which was unquestionably a water-epidemic, speaks for this theory. It may be objected, on the other hand, that the cholera-epidemic on board the two ships in the Hamburg harbour, though in them too the medium of infection was water, was of a mild character.

It would be natural to conjecture differences in the virulence of the infectious matter itself. But all investigations with a view to deciding this question have failed. As regards virulence and the power of producing the poison peculiar to cholera, no constant differences have hitherto been discovered between the cholera-bacteria of the mildest cases of the after-epidemic and those of the severest cases of the beginning of an epidemic.

I must confess that I cannot yet give any satisfactory explanation of this striking difference in the behaviour of cholera, which shows itself both in small groups of cases and in whole epidemics; and I look upon the clearing up of this mysterious phenomenon as one of the most important tasks in the further investigation of cholera.

In Altona it fortunately happened that the causes of the epidemic were early recognised, and could be removed without delay. The authorities, who perfectly understood the situation, at once did all that was necessary in order to get the upper hand of the pestilence. The disturbance in the working of the filters was most promptly remedied, the infected well was closed, all cholera-cases in which there was reason to apprehend a transmission of the infectious matter were isolated, and, when the state and situation of dwellings were too unfavourable, the inmates exposed to infection were removed to other quarters, while at the same time judicious steps were everywhere taken with a view to disinfection. The fact that the epidemic died out quickly and without great loss of life is undoubtedly due to this wisely directed action.

The last cholera-case in Altona was observed on the 12th of February 1893. Nothing of the kind has since happened, and one may therefore regard the Altona epidemic as utterly extinct.

III. THE CHOLERA-EPIDEMIC IN THE LUNATIC-ASYLUM AT NIETLEBEN NEAR HALLE.

Though it has not been possible to prove the immediate connection between the cholera-epidemic at Nietleben and that of Hamburg, the former must nevertheless be regarded as an offshoot of the Hamburg after-epidemic. In the middle of January 1893, when cholera broke out at Nietleben, it still prevailed only in Russia, France, and Hamburg-Altona. Its introduction into Nietleben from abroad is, owing to the seclusion of the asylum, simply out of the question. Unless therefore one prefers to assume that the infection was introduced in summer from one of the places then infected, and remained latent till winter, which is inadmissible for reasons to be discussed below, no alternative remains but to attribute the epidemic to a fresh but unrecognised introduction from Hamburg.

In the neighbourhood of Halle (see diagram 2), and for a good many miles below it, the Saale has a very winding course, and repeatedly divides itself, especially, for instance, at Halle, into several arms. On both sides of the river, in this part of its course, one finds porphyry, at some points hardly rising above the level of the valley, at others forming knolls, sloping steeply towards the river. A large part of the city of Halle stands on porphyry, and the provincial lunatic-asylum at Nietleben, which is situated on the opposite western side of the valley, at a distance of four kilometres from Halle, also stands on such a porphyry knoll. At the foot of the latter flows an arm of the Saale, called "the Wild Saale." The Nietleben porphyry knoll rises 30 metres above the mean level of the Saale. It is connected with the almost equally high ground behind it by a narrow ridge, corresponding in the plan (diagram 3) to the road which runs from building 18 westwards past a barn. The knoll slopes towards the two sides where the sewage-fields are marked in the plan; towards the river it ends in a pretty steep slope, at the foot of which lies the pumping-station. On this latter side it is not bounded by a regular curve, but has two short spurs, on one of which (going almost due eastwards) stand pavilions 29, 30, and 31, on the other (going southwards) pavilions 32, 33, and 34. Between these two spurs a steep-banked hollow bites, reaching to the building marked

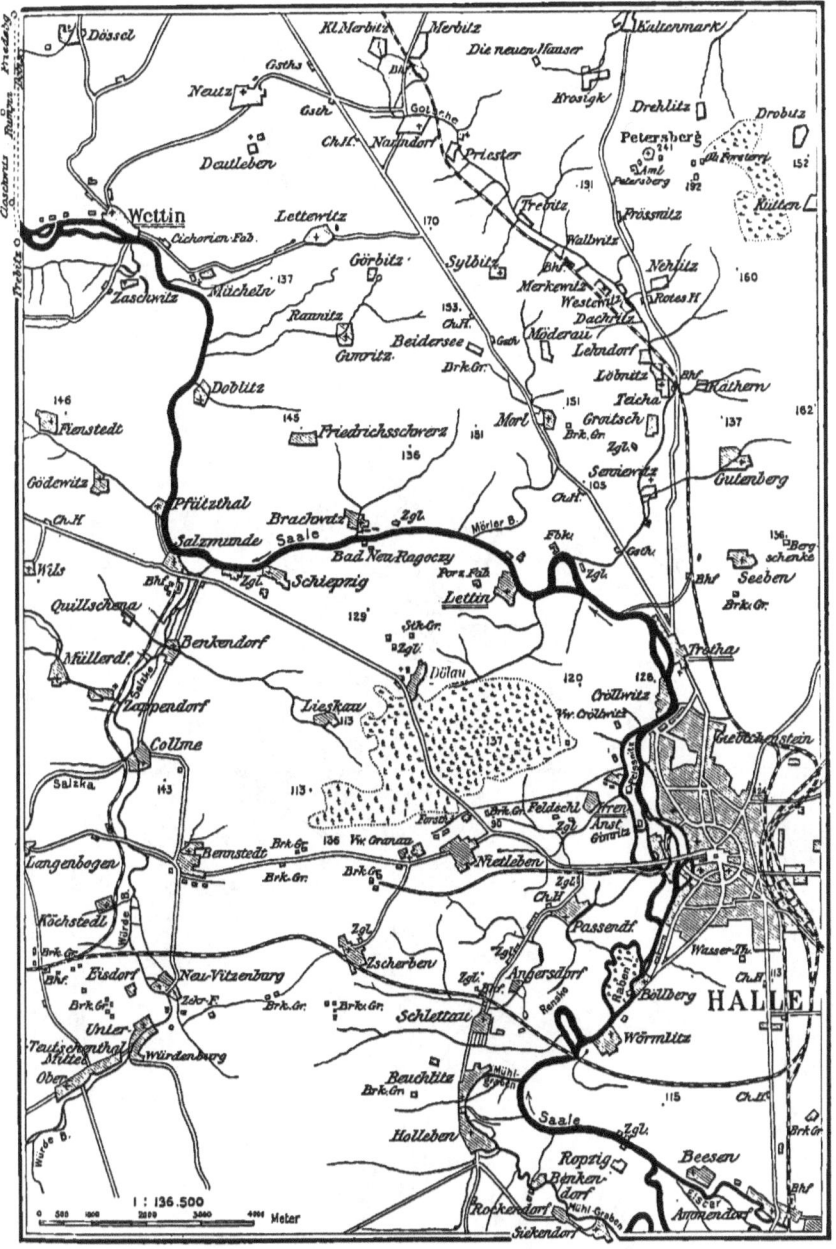

FIG. 2.—Map of Halle and its Environs (the Cholera places are underlined).

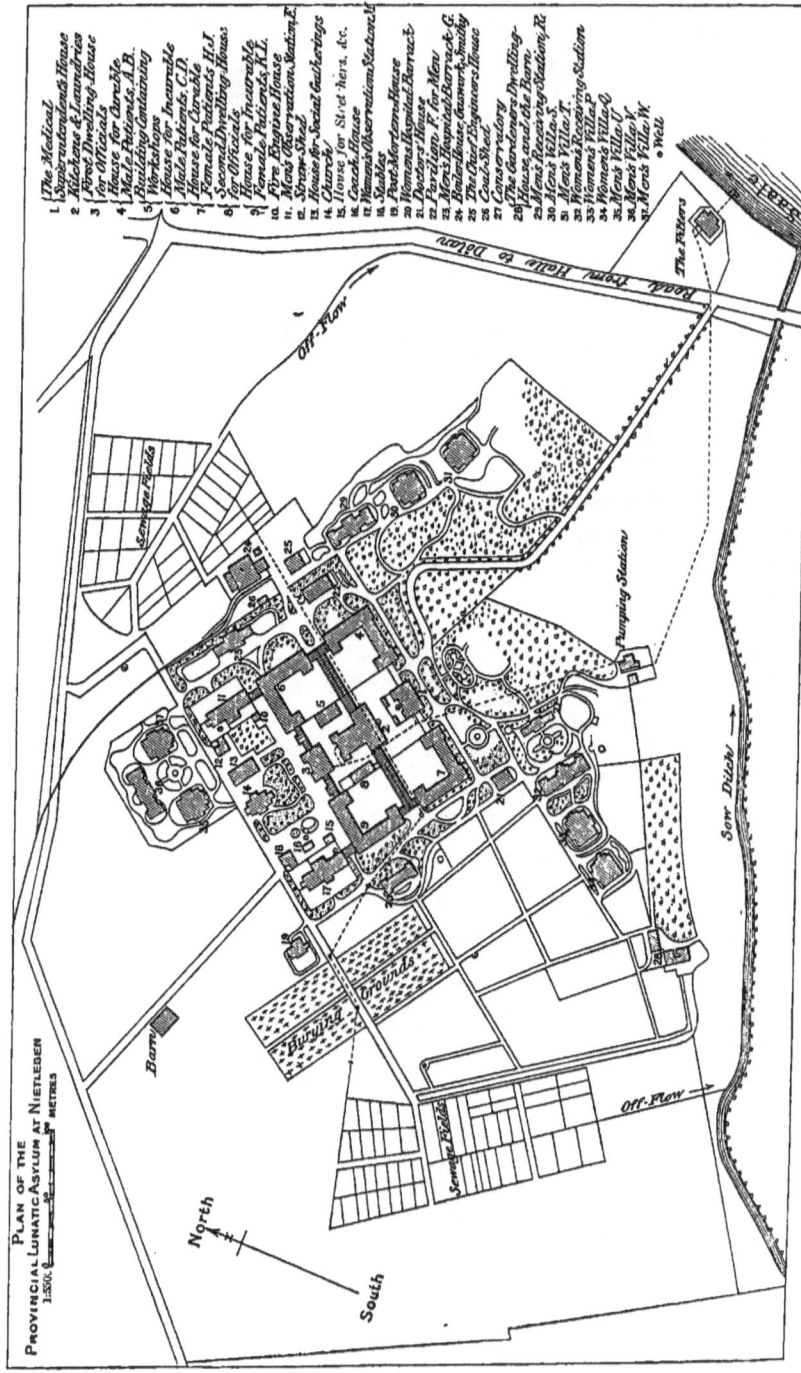

The Cholera in Germany during the Winter of 1892-93. 105

1 in diagram 3. At some points the naked rock is visible, but, with these exceptions, the surface of the porphyry is disintegrated and covered with the products of decomposition, consisting in loose fragments of stone and a loamy cement. This layer of the products of disintegration was of consider-

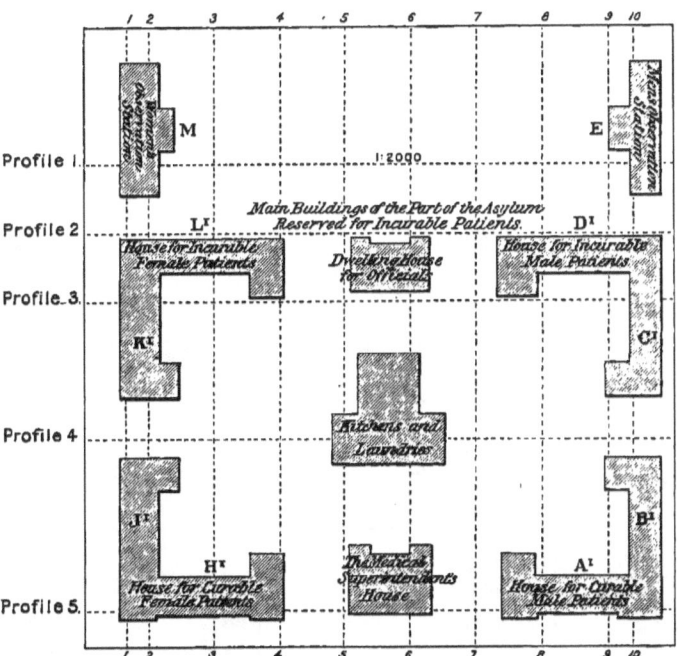

FIG. 4.—Plan shewing the Position of the Profiles in Diagram 5.

able thickness only on the top of the knoll. In the rocky substratum there a groove-like hollow has formed, beginning near building 3, running under buildings 2 and 1 towards the above-mentioned hollow, gradually getting deeper and deeper, and ending in the hollow. The situation and extent of this groove may be best seen from diagrams 4 and 5, which represent some cross-profiles of the top of the porphyry knoll. Whether this groove existed from the first, or was formed in course of time by more rapid decomposition, it obviously furnishes the natural drainage for the broadly vaulted, and in the middle somewhat sunk, surface of the porphyry knoll. Further information regarding the shape of the hill may be got from diagrams 7a and 7b, which

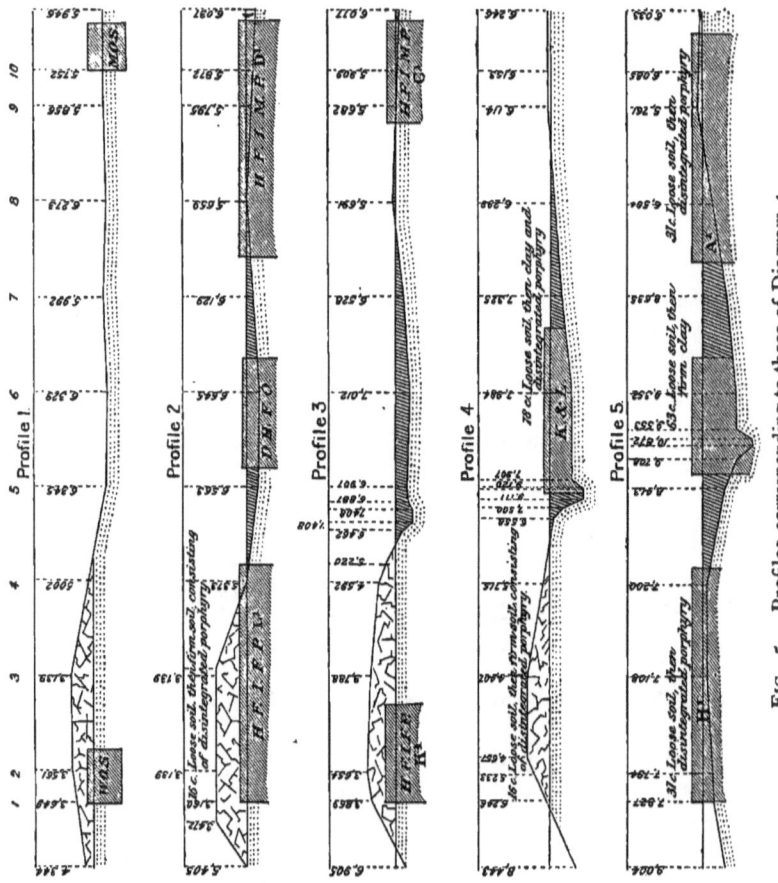

FIG. 5.—Profiles corresponding to those of Diagram 4.

show the elevations in lines drawn through the front of the asylum-buildings and through the longitudinal and transverse directions of the two spurs of the hill. The situation of these profile-lines is easily discoverable with the aid of the corresponding letters in diagram 6. The structure of the ground on which the asylum stands must, accordingly, be conceived as follows: The nucleus of the hill consists of compact porphyry, disintegrated and loosened in its outermost layer, which is one to two metres thick. This rind has clefts and rents, filled with loamy masses.* Over it lies a layer of loose stones and loam of varying thickness, and over that arable soil. The rock slopes in all directions, so that there is nowhere any opportunity for underground water to gather. Even on the top, where there is a shallow depression, no underground water can accumulate, for there is nothing to prevent the water that oozes in from above from flowing off through the groove already described.† It is equally impossible for water to stream into the precincts of the asylum from the higher land to the west, for the connecting ridge is too narrow for that, and is furnished on both sides, almost up to the top, with artificial drainage, which diverts any underground water that may stream in the direction of the asylum towards the sewage-fields. The existence of underground water might, indeed, be inferred from the fact that there are several wells between the buildings of the establishment on the top of the hill; but these wells are driven deep down into the compact rock, and form, as it were, cisterns, which collect a certain quantity of water trickling in from above, but yield no real underground water.

All the buildings of the establishment stand on the top of the porphyry knoll, and are founded on the firm undisintegrated rock, which had to be prepared for them at some places, especially under pavilions 29, 30, 31, 32, 33, 34, by blasting. For this reason the cellars of the buildings, as I have convinced myself by personal inspection, show no traces of

* The volume of the pores of the disintegrated porphyry fluctuates between 5 and 6·5 per cent. It rises, in proportion to the degree of disintegration, to between 8·5 and 11·5 per cent. The highest percentages that have been obtained are 11·6 and 11·7.

† The groove is transversely blocked near its outflow by the medical superintendent's house, but care has been taken that water flowing down the groove may find its way out past that building through a deep drain laid between it and the edifice in which the curable male patients live.

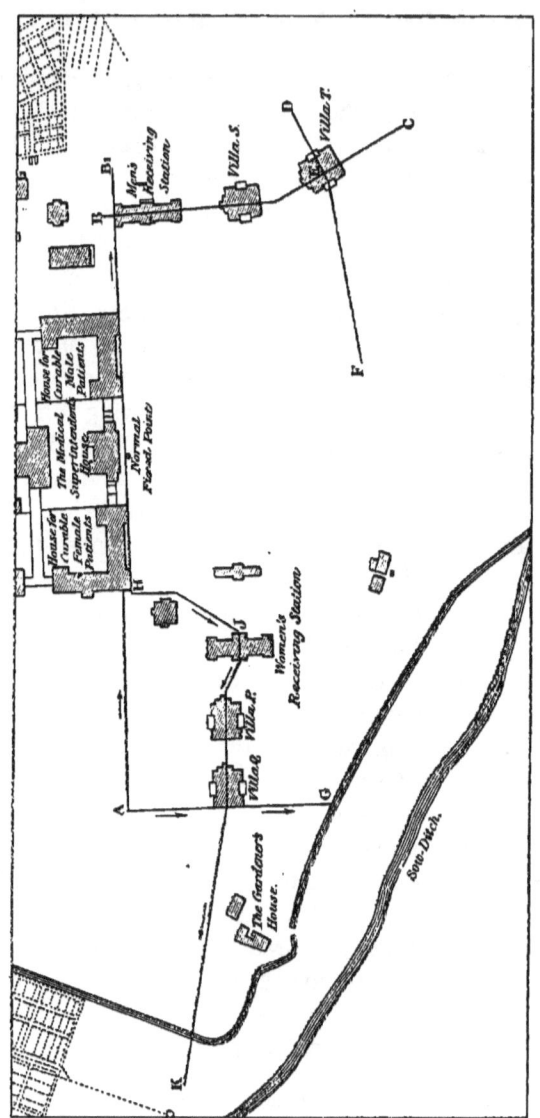

Fig. 6.—Plan shewing the Positions of the Profiles in Diagrams 7a and 7b.

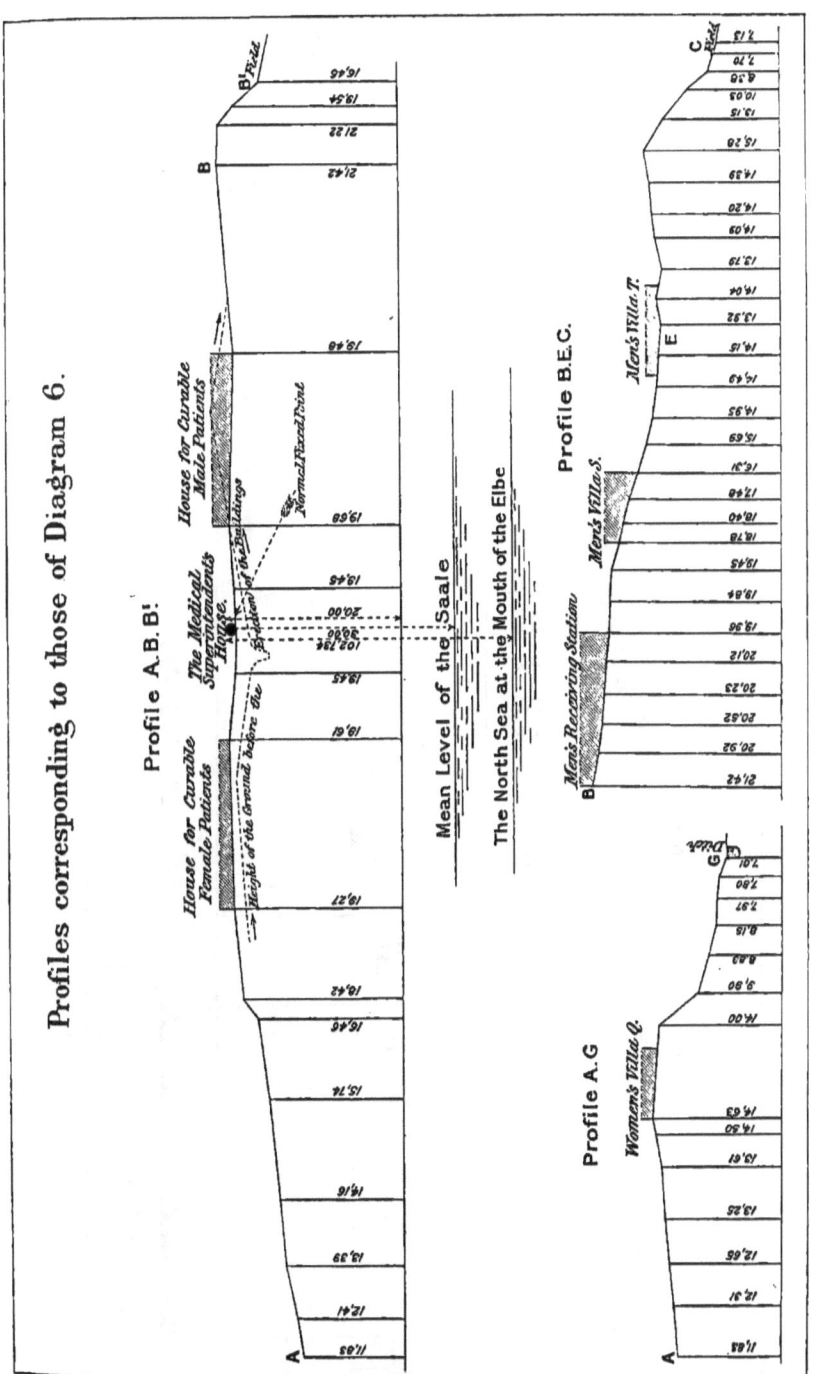

FIG. 7a.—Profiles corresponding to those of Diagram 6.

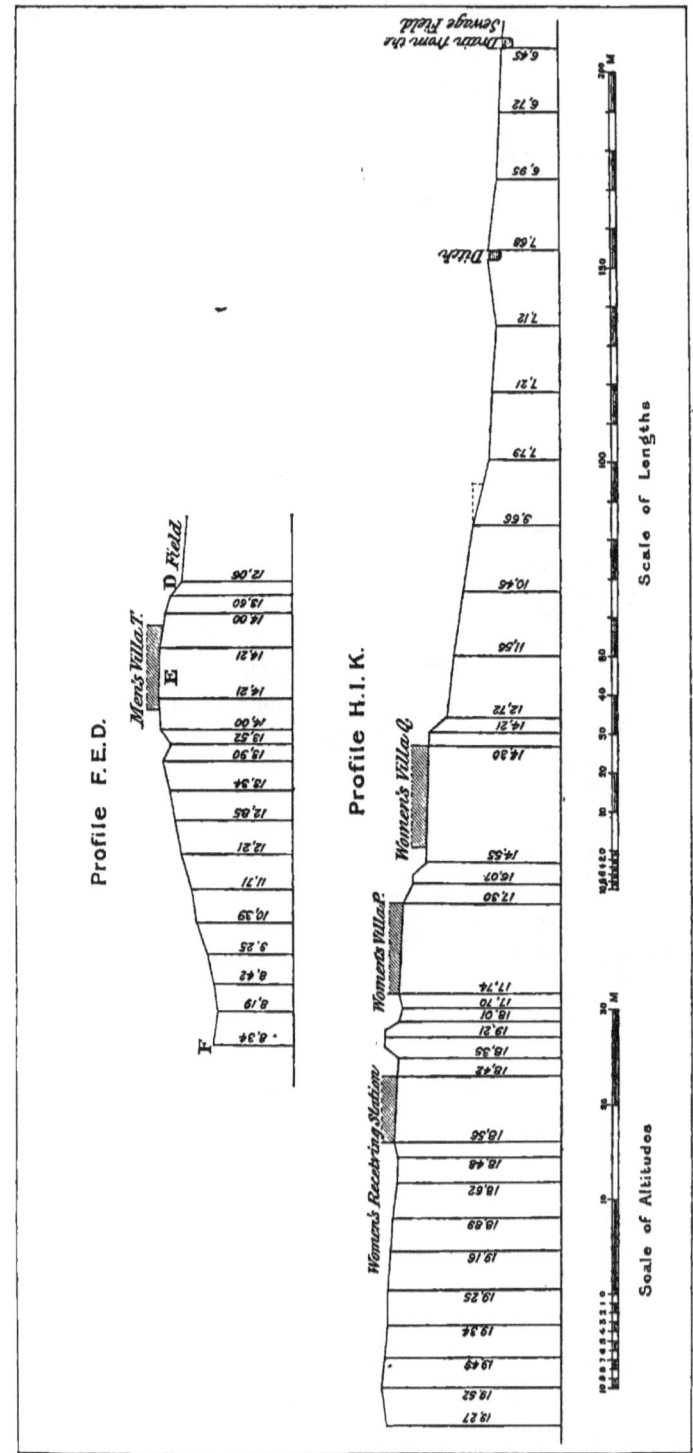

FIG. 7b.—Profiles corresponding to those of Diagram 6.

dampness emanating from the soil. Wherever such traces seemed to exist, under building 6 for instance (for incurable male patients), it was ascertained that they were caused by water from the waste-pipes of the steam-heating apparatus, which had become leaky. The only inhabited building belonging to the establishment not founded on rock is the gardener's house (28). It lies at the foot of the hill on alluvial soil, in which the underground water rises and falls. At high level the water is said partly to flood the cellars of the house.

The asylum was built in 1840, and originally consisted of a group of edifices in rectangular order, comprising buildings 1, 2, and 3, containing offices and dwellings for persons engaged in the management, and the two-storied corridor-buildings 4, 6, and 7. A similar corridor-building (9) for incurable female patients was afterwards added, and several groups of pavilions were built still later outside of the rectangle, namely, pavilions 29, 30, and 31 on the men's side, 32, 33, and 34 on the women's side, and 35, 36, and 37 west of the main buildings.

The establishment was provided with water-pipes and sewers nearly ten years ago. The construction and management of the water-system, which receives its water from the Wild Saale, has been described in detail in a previous article on water-filtration and cholera. Its main course is indicated in the plan. Close to the Wild Saale lie the filters, from which an underground iron pipe leads the water to the pumping-station at the foot of the hill. The pump then raises it and presses it into high reservoirs in buildings 2 and 3, whence it is distributed all over the establishment by pipes. Another system of pipes collects the dirty water from the sinks, water-closets, bath-rooms, kitchens, etc., and leads it into two main sewers, which begin at building 2, and go in opposite directions to the sewage-fields on the slopes of the hill on both sides of the establishment. These sewers are of mason-work, are provided with man-holes, and are up to date in every respect. The sewage-fields too are properly laid out.

They consist of horizontal beds rising in terraces. Their soil is deeply enough drained. The off-flowing water is collected in a main conduit, and led in open ditches, the course of which is indicated in the plan, both from the northern and from the southern sewage-field, into a dirty little brook called the Saugraben (Sow-Ditch). That this brook flows into the Wild Saale at a very short distance above the place where the water for the asylum is taken from the river, and to what grave objections such an arrangement of the sewerage and the water-system is open, I have already stated in the article referred to above.

The asylum was originally intended for 600 patients, but the number has gradually risen to 800, and, if we count the attendants, the doctors, the officials, and the other persons employed in the establishment, the number of its inmates is now about 1000.*

The Nietleben asylum was repeatedly the scene of cholera-epidemics in former years, and that under circumstances which seemed quite specially favourable to the application of the soil-theory. These epidemics therefore play a certain part in the literature of cholera, are frequently referred to, and cannot well be left out of account here. We owe our information regarding them to Delbrück,† from whose writings on the subject I take the following items :—

In 1866 the cholera was restricted to the men's department, where 17 persons died during the epidemic, which lasted 36 days. Opportunities of transmission to the women's department were not wanting; and diarrhœa prevailed there during the epidemic, but no fully developed cholera-cases occurred. The contrary happened in 1850, when the men's department remained free, whereas the women's suffered severely. Del-

---

* On the 14th of January 1893 the asylum contained 436 male patients, 375 female patients, 124 officials who had and 56 who had not their meals in the establishment,—in all 991 souls.

† Delbruck's Report of the Cholera Epidemic of 1866 in Halle, published in Halle in 1867, page 19. The Deliberations of the Cholera-Conference at Weimar, edited by Thomas and published in Munich in 1867, page 24.

brück seeks the reason for this behaviour of cholera in the structure of the soil. The building marked 9 in the plan, and now occupied by incurable female patients, did not exist in 1850; but on the place on which it now stands there was "a big hill, which, forming a steep slope, increased the afflux of water etc., and prevented the off-flow of moisture from this place; this hill, however, was removed when the building was erected, and the water is thus enabled to flow off very easily." If one reads this description of the ground, without having seen the place, one cannot but get the impression that the soil may have had some influence in this case. But it really had none whatever. On close investigation the "big hill," with its "steep slope," shrinks into nothing, and Delbrück's description of the circumstances connected with the former cholera-epidemics at Nietleben shows once more with signal distinctness how naively inquiries into the etiology of cholera used to be gone about, and how cautious one ought to be in using the cholera-reports of former times. A glance at diagram 5, which represents the ground in question in its earlier form and after the erection of building 9 (profiles 1 to 4), at once informs one that the alleged "big hill" was nothing more than a swelling about 2 metres high at the southern edge of the ground on which the asylum stands. Of a "steep slope" nothing whatever is visible, unless indeed the above-described groove, which is situated inwards from the place in question, be meant. But just the existence of this groove should have convinced Delbrück that the off-flow of fluids takes place in quite another direction than he supposed. Moreover, the "hill" was replaced by an extensive edifice, which, with its foundation-walls resting firmly on the rock, would have presented at least as great obstacles to the motions of the underground water, supposing such motions to have taken place there, as the "hill." Fortunately, however, the archives of the asylum contain more exact memoranda of the course of that epidemic, which were most readily placed at my disposal by the medical superintendent. Dr Buchholz, the first assistant-physician of the asylum, has taken the

H

trouble to mark the cases in the ground-plans of the buildings in which they occurred, and to excerpt the necessary notices from the reports; and I may restrict myself to reproducing his sketches and notices here, as received by me from him, because they contain everything that is of interest to us here.

1. THE EPIDEMIC OF 1850 (Diagram 8).

At the time when this epidemic occurred, all the female patients were lodged in the two-storied house marked 7 in the plan. The building was intended for 75 patients, but was occupied by 110. The imbecile patients lay in the ward marked J I., the convalescents in that marked J II. The

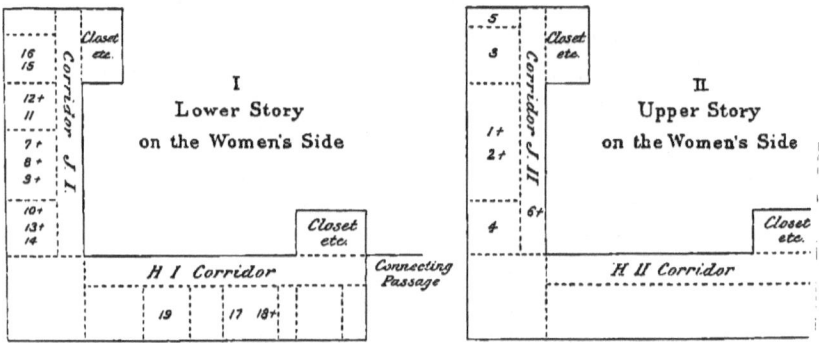

FIG. 8.—Cholera Epidemic in the year 1850.

water came through the old conduit from the Saale (without filtration), but water from the wells was also used for drinking. For the fæces there were cesspits.

With regard to the patients the following notes were taken:—

Number 1, patient T., received from Merseburg, which was infected with cholera (no cholera in the house, but cases of diarrhœa). Received on the 14th of August; fell ill in the night between the 17th and 18th of diarrhœa, which was recognised as cholera on the 18th; died on the 19th.

Number 2, an attendant in the above corridor marked J II., who had attended Number 1. Had diarrhœa on the 19th of August, soon followed by marked cholera-symptoms, and died on the 20th.

Numbers 3 and 4, patients of corridor J II., fell ill on the 23rd of August.

Number 5, a patient of corridor J II., fell ill on the 24th of August.

Number 6, the chief attendant, who had tended the patients, and sat up with them, fell ill on the 24th, and died on the 25th of August.

Till then cases had occurred only in the upper corridor, J II.; then the disease broke out in the lower corridor, J I.

Number 7, a patient, fell ill on the 24th of August, and died in the night between the 25th and the 26th.

On the 25th several patients of corridor J I. suffered from diarrhœa.

Number 8, a patient, fell ill and died on the 26th.
„ 9, „ „ on the 26th, and died on the 27th.
„ 10, „ „ „ 27th (died several weeks later of exhaustion).
„ 11, „ „ „ 27th.
„ 12, „ „ „ 27th, and died on the 28th.
„ 13, „ „ „ 28th, and died on the 29th.
„ 14, „ „ „ 28th.
„ 15, „ „ „ „
„ 16, „ „ „ „

Then the disease broke out in corridor H I.

Number 17, an imbecile child, fell ill on the 29th.

On the 29th several patients suffered from diarrhœa, one severely.

Number 18, an imbecile child of nine, in the same room as number 17, fell ill and died on the 31st.

2. THE EPIDEMIC OF 1866 (Diagram 9).

When cholera broke out the asylum contained 470 patients, and 90 officials, attendants, etc.,—in all 560 souls. Cholera had prevailed in Halle since the middle of July. Whether the infection was brought from there or from some other place, could not be ascertained.

With reference to the undersoil it is stated that a drain ran along the front of the asylum-buildings (marked 4 and 6 in

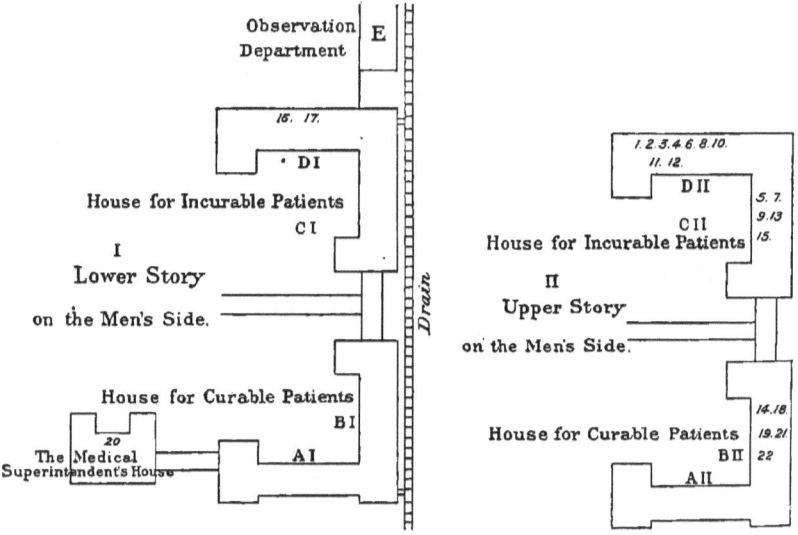

FIG. 9.—Cholera Epidemic in the year 1866.

the plan), which drain received the waste-water of the male patients' houses and the observation-station, and led it into an open ditch. This drain (six inches square) was found to be choked and grown through by roots, and water had accumulated in the cellars under the house for incurable patients in consequence. After the epidemic it was replaced by a larger drain.

The water-supply and the closets with cesspits were just as they had been in 1850. The cholera-cases occurred in the following order:—

19 August	Number	1 in	D II.
21	"	"	2 " "
"	"	"	3 " "
"	"	"	4 " "
22	"	"	5 " C II.
23	"	"	6 " D II.
24	"	"	7 " C II.
25	"	"	8 " D II.
"	"	"	9 " C II.
26	"	"	10 " D II.
27	"	"	11 " "
28	"	"	12 " "
29	"	"	13 " "
30	"	"	14 " B II. (a male attendant).
31	"	"	15 " C II.
4 September		"	16 " D I.
5	"	"	17 " "
6	"	"	18 " B II.
11	"	"	19 " "
19	"	"	20 { in the medical superintendent's house.
20	"	"	21 in B II.
26	"	"	22 " "

Of these 22 cases, 18 proved fatal; the victims were the medical superintendent, 1 male attendant, and 16 patients.

Besides the cases enumerated here, 45 persons are said to have suffered from violent diarrhœa or vomiting in the most various departments. Some 50 such cases occurred in the women's department, but the symptoms did not become alarming in a single case.

This is all that I have been able to learn about the former epidemics, but it fully suffices to afford a clear conception of the behaviour of the cholera in 1850 and 1866. Had the cholera been influenced on those occasions by general causes, it must have broken out simultaneously and equably over the whole establishment, or at least in the building subject to such influence. If, for instance, in 1866 the choked drain had been connected with the causes of the epidemic, the disease must have appeared first and predominantly in the lower story, and especially in wards B and C. But this was by no means the case. It began in the upper story in ward D, passed to the adjacent ward C II., and then to ward B II., adjacent to that. Only by the way, and almost two weeks after the outbreak of the epidemic, did two cases occur in the lower story in ward D.

If one follows the spread of the cholera on the ground-plans, and with the aid of the numbers in both epidemics, it appears at once, and quite unmistakably, that on both occasions the disease spread from one point, to which it must have been brought (in 1855 the bringing was proved), to the adjacent parts of the establishment; it crept on like a fire in the direction in which it found something to devour. The epidemics, then, were not due at all to causes lying outside of the patients, the alleged "hill" or the "choked drain"; but the patients themselves offered so favourable points of attack to cholera that it needed no special medium, and could spring directly from one person to another. This is the form of cholera which I have described at the beginning of this article as the second type, the same as we saw in the after-epidemic in Hamburg, and as has been observed frequently enough before on board emigrants' ships, in prisons, and especially in lunatic-asylums, in short, wherever human beings live densely crowded together under unfavourable circumstances. This is also the simplest explanation of the fact that, in both the former epidemics in the Nietleben asylum, the great majority of those who were attacked by cholera were incurables, that is, unclean patients.

It is extremely instructive to see how in the same establishment, which has already had two cholera-epidemics of the second type, the first or explosive type can also break out, if the circumstances fulfil the necessary conditions. For the epidemic of 1893, which I shall now describe, was of a markedly explosive character.

3. THE EPIDEMIC OF 1893.

During the summer of 1892, when the cholera was carried in all directions from Hamburg, no cases of genuine cholera were observed in Halle and its vicinity. Only one case of illness was regarded as suspicious, that of an assistant-stoker who fell ill of diarrhœa with vomiting and cramp in the calves in Nietleben Asylum on the 25th of August, and was taken at once to the university-clinic in Halle. After that nothing more of the kind occurred. Simple cases of diarrhœa, which are always more or less frequent in lunatic-asylums, were of course not wanting at that time at Nietleben either, but in the beginning of October their number increased to such an extent that 73 such cases were entered in the sick-list between the 2nd and the 27th of that month. This did not last long, however, as the following sick-list, containing only the cases of diarrhœa, dysentery, and enteric fever, shows:—

SICK-LIST.

Date.	Diarrhœa.	Dysentery.	Enteric Fever.
October 3	1	1	...
,, 10	66	...	...
,, 26	6	...	...
November 12	...	...	2
,, 17	1	...	...
,, 24	2	...	...
,, 25	2	...	...
,, 26	1	...	...
December 4	1	...	...
,, 8	1	...	...
,, 11	2	...	...
,, 16	4	...	...
,, 20	...	1	...
,, 23	3	...	..
,, 26	3	...	2
January 2	3	...	...
,, 7	1	...	1
,, 8	...	...	1

In November and December but few cases of diarrhœa occurred, but nine cases are noted in the last days of December and the first two of January. Between the 2nd and the 15th of January only one case occurred. There was accordingly no premonition of the fatal catastrophe that was about to befall the establishment. I wish to draw special attention to this point, as it has been repeatedly asserted that the Nietleben epidemic was preceded by premonitory cases of diarrhœa. The cases which occurred in October, a quarter of a year before, cannot possibly be designated as premonitory; apart from them, however, the time before the epidemic, and especially the weeks immediately before it, were marked by nothing that could be inferred to be connected with cholera.

Speaking generally, I have not yet been able to convince myself that the so-called genius epidemicus reveals itself in such a manner before a cholera-epidemic. In places where cholera has unexpectedly broken out, one finds on unbiased inquiry no striking increase of disturbances of digestion, or at most only such an increase as occurs at the same season in other years also. This was the case in Hamburg before the great epidemic and at Nietleben. The supposed premonitory cases of diarrhœa, on the other hand, are always found where an outbreak of cholera is expected with eager attention; there every case of diarrhœa, vomiting, etc., no matter how insignificant, is taken notice of, and, if possible, ascribed to cholera. After the outbreak of cholera in Hamburg, for instance, the patients sent to the Berlin hospitals on suspicion of cholera were strikingly numerous; a somewhat imaginative observer might unquestionably have seen the hand of the genius epidemicus in that. In reality, however, they were the usual cases of summer diarrhœa, indigestion, alcoholic intoxication, etc., cases which, but for the fear of cholera, would not have been sent to the hospitals in such numbers at all.

In Berlin, then, there were premonitory cases of diarrhœa in sufficient number, yet no cholera followed; in Hamburg and Nietleben, on the other hand, where cholera broke out unexpectedly, they were wanting, and the much talked-of genius epidemicus showed no trace of himself just where his warning influence would have been quite specially in place.

The first case of cholera was observed at Nietleben on the
14th of January 1893. One of the patients of the asylum

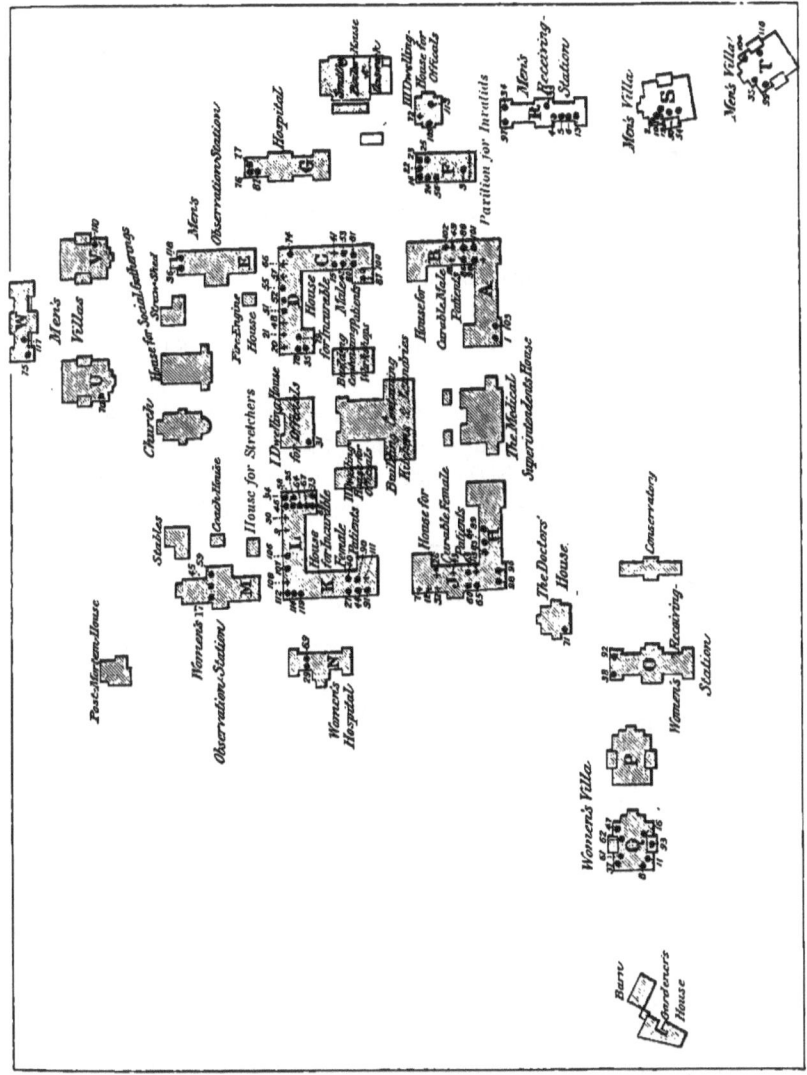

Fig. 10.—The Distribution of the Cholera Cases in the Lunatic Asylum at Nietleben. + Indicates the Cases that occurred in the upper story.

fell ill quite suddenly of violent diarrhœa with vomiting, and died on the same day. The clinical symptoms were those of Asiatic cholera, the results of the autopsy agreed with them, and the cholera-bacteria were found in the contents of the intestines of the corpse.

This first case, which happened in the building marked A in diagram 10 (4 in diagram 3), was followed next day (the 15th) by six, all of which proved fatal, and by eleven more on the 16th, eight of which proved fatal.

In direct contrast to the epidemics of 1850 and 1866, in which the cholera had begun at a definite point and only gradually crept on to the adjacent rooms and wards, it broke out this time suddenly in the most different parts of the establishment, both on the men's and on the women's side. The eighteen cases of the first three days occurred in eleven different wards and ten different buildings of the asylum. Diagram 10, in which they are all entered, shows the local distribution of the cases, and the following table shows their sequence in point of time :—

Date.	Cases.	Deaths.	Date.	Cases.	Deaths.
January 14	1	1	January 25	5	4
,, 15	6	6	,, 26	3	1
,, 16	11	8	,, 28	2	1
,, 17	15	7	,, 31	1	1
,, 18	8	2	February 1	1	...
,, 19	7	2	,, 4	1	1
,, 20	16	6	,, 5	1	1
,, 21	9	3	,, 10	1	1
,, 22	12	5	,, 13	1	...
,, 23	8	1			
,, 24	13	1	Total	122	52

Of the 122 cholera-patients 63 were men (including three doctors), 59 women (including seven attendants and three wives of officials).

In order to give a more exact view of the local and temporal distribution of the cholera, I add the following table, which names the various departments of the establishment, and states the number of beds in each department, the number of its inmates on the day when the cholera broke out, and of the cases that occurred in it, with the dates of their occurrence:—

I. THE MEN'S SIDE.

Department.	Number of Beds.	Number of Inmates on 14th Jan. 1893, including the Attendants.	Cholera-Cases.	Remarks.
Curable male patients' house, lower story (A I.).	30	28	2 (Jan. 14, 24).	
Curable male patients' house, upper story (A II.).	9	8	0	Seven patients of the first and the second class, with one attendant.
Curable male patients' house, lower story (B I.).	34	34	4 (Jan. 23, 24, 24, 24).	
Curable male patients' house, upper story (B II.).	34	33	4 (Jan. 16, 20, 23, 25).	
Incurable male patients' house, lower story (C I.).	20	18	4 (Jan. 16, 20, 23, Feb. 10).	Sixteen patients of the first and the second class, with two attendants.
Incurable male patients' house, upper story (C II.).	51	48	5 (Jan. 19, 19, 22, 22, 23).	
Incurable male patients' house, lower story (D I.).	54	52	5 (Jan. 20, 20, 22, 22, 23).	
Incurable male patients' house, upper story (D II.).	46	45	7 (Jan. 17, 17, 20, 20, 20, 21, 22).	
Men's observation-station (E).	30	29	2 (Jan. 18, Feb. 4).	
Men's pavilion (F).	33	32	7 (Jan. 15, 16, 17, 17, 17, 17, 20).	
Men's hospital (G).	36	23	2 (Jan. 22, 22).	
Men's receiving-station (R).	33	32	7 (Jan. 15, 15, 15, 16, 18, 18, 24).	
Men's pavilion (S).	44	37	5 (Jan. 15, 16, 16, 20, 20).	
Men's pavilion (T).	44	36	4 (Jan. 18, 24, 24, 26).	
Men's pavilion (U).	44	0	1 (Jan. 21), a female attendant.	Used immediately after the outbreak of the cholera as a cholera-hospital. The patients in V were removed to other departments.
Men's pavilion (V).	44	36	1 (Jan. 26), an incurable patient employed in the disinfecting-house.	
Men's pavilion (W).	44	0	3 (Jan. 22, Feb. 1, 13), three female attendants.	

124 *The Cholera in Germany during the Winter of* 1892-93.

II. THE WOMEN'S SIDE.

Department.	Number of Beds.	Number of Inmates on 14th Jan. 1893, including the Attendants.	Cholera-Cases.	Remarks.
Curable female patients' house, lower story (H I.).	34	31	6 (Jan. 17, 22, 22, 23, 23, 24).	The woman who fell ill on the 24th was an attendant.
Curable female patients' house, upper story (H II.).	12	9	0	Eight patients of the first and the second class, with one attendant.
Curable female patients' house, lower story (J I.).	32	31	3 (Jan. 20, 20, 21).	
Curable female patients' house, upper story (J II.).	39	38	6 (Jan. 15, 16, 17, 17, 19, 25).	
Incurable female patients' house, lower story (K I.).	43	40	4 (Jan. 17, 18, 19, 23).	
Incurable female patients' house, upper story (K II.).	28	28	1 (Jan. 26).	Twenty-four patients of the first and the second class, with four attendants.
Incurable female patients' house, lower story (L I.).	47	44	12 (Jan. 19, 20, 21, 21, 21, 22, 24, 25, 25, 28, 31, Feb. 5).	
Incurable female patients' house, upper story (L II.).	42	41	5 (Jan. 16, 17, 18, 24, 25).	The woman who fell ill on the 17th was the upper attendant.
Women's observation-station (M).	33	33	3 (Jan. 16, 19, 20).	
Women's hospital (N).	24	25	2 (Jan. 17, 21).	
Women's receiving-station (O).	33	34	2 (Jan. 18, 23).	
Women's pavilion (P).	44	23	0	
Women's pavilion (Q).	41	42	8 (Jan. 16, 16, 16, 18, 19, 20, 24).	The woman who fell ill on the 24th was an attendant.

III. THE OFFICIALS' HOUSES.

Department.	Cholera-Cases.	Remarks.
First officials' house.	2 (Jan. 17, 17).	Wives of officials.
Third officials' house.	3 (Jan. 21, 24, 28).	Two doctors and the wife of an official.
The doctors' house.	1 (Jan. 21).	A doctor.

These tables and diagram 10 show without explanation that, with the exception of some parts, to which I shall return, the cholera was pretty equally distributed over the whole establishment, not only at first but also during the further course of the epidemic. The causes must therefore have acted not on certain buildings or certain groups of the inmates but on the whole establishment. In the case in question only the soil, the provisions, or the water could have exercised such a common influence.

The soil was quite out of the question from the first. All the buildings stand on solid rock. The only ones that could possibly have been unfavourably influenced by the soil were the older edifices, which form a rectangle, and are grouped round the depression on the top of the porphyry knoll (A, B, C, D, H, J, K, L, in diagram 10). Owing to the natural and artificial drainage of the depression, a swelling of the underground water seemed out of the question; but the loose soil with which the depression is filled was probably impregnated with sewage in former times, when the liquid filth of the establishment was not yet removed by sewers, and one might have expected that this polluted soil would influence the behaviour of the cholera in some way or other. Of this, however, there was not the slightest trace, for the cholera did not behave a whit otherwise in the older buildings than in the new ones, the rocky undersoil of which had never been exposed to unusual contamination.

As regards the food-supply of the establishment also, careful inquiry failed to discover any ground for supposing that the general infection had been transmitted by articles of nourishment. Most of these articles were supplied at the same time by the purveyors to the clinical institutions in Halle, without their being attacked by cholera; and among the cholera-patients in the asylum there were persons whose meals were not provided by the establishment.

The only possible explanation left was that the water had been the bearer of the infectious matter. From the first, indeed, there were weighty reasons against this assumption, for, just in this respect, the asylum was provided with apparatus which ought to have afforded adequate protection.* Supposing that the infectious matter of cholera had by some accident or other been brought into the establishment, and had got into its sewage, it must have been retained by the filtering effect of the soil in the sewage-fields; and, even if the sewage-fields had let it escape, it must have remained on the surface of the sand-filters through which the water passed before entering the asylum. That the infectious matter should have surmounted these two obstacles, both of which, as experience taught, were able to keep it away, was not very probable. Nevertheless it was necessary to examine the filter-work and the sewage-fields, in order to ascertain whether they performed their functions in such a manner that they were really able to keep off the infectious matter.

I have already minutely stated the results of the investigation of the filter-work in my article entitled, "Water-Filtration and Cholera." It showed, to repeat the contents of that statement here in brief, that the work had indeed some faults of construction, but could under careful management have supplied water free of infectious matter. The management, however, was such that the water passed through the sand-filters almost unfiltered. Here, then, was a big gap in the sanitary arrangements for the protection of the asylum against infection.

The function of the sewage-fields was not much better performed. They are, on the whole, properly laid out, but have no basins to retain the dirty water in winter, when the ground is frozen and impervious, till the frost is over. The reason why no such basins were made when the sewage-fields were laid out probably was that, in case of need, the low-walled beds could serve in their stead. In this case one bed after the other must have been filled to its utmost capacity with dirty water. The surface of the beds is large enough to

* The wells of the establishment had not to be taken into account, for they had been closed for half a year past.

admit of the dirty water's accumulating on the sewage-fields during a long period of frost, and the fields could have performed their function properly even in the hard winter of 1892-93. But I regret to have to say in this case too, just as in the case of the filter-work, that none of the persons concerned seem to have been acquainted with the proper treatment of sewage-fields; for no attempt was made to accumulate the dirty water on the fields during the frost. The consequence was that, as soon as the ground was frozen, the dirty water flowed over the surface, or in the larger clefts and mouse-holes of the fields, without getting any cleansing worth speaking of from the soil. When I examined the sewage-fields, they were covered with a pretty thick sheet of snow, under which the soil was frozen almost a metre deep; and under the snow, but over the frozen ground, the dirty water flowed off without let or hindrance wherever it had made itself a way. The water which flowed to the sewage-fields and that which flowed from them were repeatedly subjected to bacteriological investigation, and, as was to be expected under such circumstances, were found not to differ essentially from each other. In one case, for instance, dirty water taken before it reached the sewage-land on the women's side was found to contain 400,000 germs per cubic centimetre water taken at the same time from one of the beds 350,000, from a side-ditch 450,000, from the main drain-pipe at the lower end of the sewage-land, 470,000.

This protective arrangement, accordingly, also proved utterly insufficient, and the infectious matter was free to circulate through the establishment with the stream of fluids. That this actually happened was irrefutably proved by the further bacteriological investigation, for the cholera-bacteria were found at various points in this circulating stream. They were found on the women's side in the dirty water at the place where it enters the sewage-land, in the sewage-fields themselves, and in the water which left the sewage-land through the main drain-pipe, and on the men's side too at the places where the water enters and leaves the sewage-land.

They were found in the water of the Wild Saale also below the mouth of the Saugraben (Sow ditch), in the filtered water of filter number II., and in a specimen of water taken from one of the water-pipes in the asylum.

The fact that the proof of the presence of the cholera-bacteria succeeded so completely is unquestionably due in this case, as in that of the cholera-well in Altona, to the circumstance that the investigation was made so soon after the outbreak of the epidemic.

The immediate cause of the explosive outbreak of cholera at Nietleben is thus rendered perfectly clear. The infectious matter must have been brought into the asylum in some way or other, then passed with the dirty water of the establishment over the frozen sewage-fields into the Saugraben, from the Saugraben into the Wild Saale, and from the Wild Saale through the water-conduit back to the asylum. The water was accessible to all the inhabitants of the establishment, and an explosive epidemic, attacking the whole asylum more or less equally according to the quantity of cholera-bacteria in the water, the relations of the inmates to the water, and their individual predisposition, was inevitable, and accordingly took place.

In this case also the water cannot have remained the only source of infection. Among people so uncommonly susceptible as the inmates of a lunatic-asylum are secondary infections undoubtedly took place. As such I am disposed to regard some of the cases that occurred in the wards for incurables with their unclean patients, especially, for instance, in ward L I., where more than one-fourth of the inmates were attacked. The female attendants too who got the disease while engaged in tending cholera-patients in villas U and W, which were used as a cholera-hospital, were probably not infected by the water, but caught the infection directly from the patients. It is also probable that the three doctors who were attacked were infected in the same way.

As the cholera-cases were most numerous in the parts of the establishment where opportunity was given for secondary infections, so, on the other hand, they were most rare where the circumstances were less favourable to the transmission of infection. This was the case in wards A II., H II., and K II., in which first-class and second-class patients were kept. A II. with eight and H II. with nine inmates were entirely spared, and in K II., with its twenty-eight inmates, only one case occurred. The explanation of this phenomenon probably is that the first-class and second-class patients get more coffee, tea, and other beverages than those of the third class, and that they consequently came less into contact with the infected water. This explanation is contradicted, indeed, by the fact that four cases occurred among eighteen persons in ward C I., which is also occupied by first-class and second-class patients. Though one of these four cholera-patients, the last who was attacked, had been removed from D I. to C I. shortly before, and had probably caught the infection in D I., there still remain three cases, a strikingly large number, and I have not succeeded in getting a satisfactory explanation of this.

It is also a singular fact that the women's pavilion P was entirely spared. Neither its construction nor its undersoil differs from those of the adjacent women's pavilion Q, and of the corresponding men's pavilion S, both of which had numerous cases. The only distinguishing circumstance was that pavilion P was only half occupied when the cholera broke out, and its exemption was probably due to a certain extent to this. But, whether this alone, or in combination with the circumstance that this pavilion had a specially careful attendant, who is said to have given her charges boiled water only, suffices to explain the phenomenon, is a question which I cannot decide.

It remains to be mentioned that no case of cholera occurred in the gardener's house, which had five inmates. It is the only inhabited house of the establishment which stands not on rock but on the alluvium of the valley, which is richly contaminated at this spot and subject to changes of moisture. According to the soil-theory it offered a particularly favourable point of attack to cholera. The members of the gardener's family get their water from a pipe of the water-conduit, which pipe comes nearly to the house. They admitted too that they had drunk the water, but only in small quantities and after dinner. I am disposed, however, to attach less weight to this than to the small number of the inhabitants of the house. This circumstance may have saved them as well as the first-class and second-class wards, A II. and H II.

After the investigation had proved to a certainty that the explosive Nietleben cholera-epidemic had been caused by the infecting of the water of the establishment, it was necessary to inquire further how the infectious matter could have got into the water-conduit. I have already indicated my opinion as to the immediate origin of the infectious matter, and that I suppose the infecting of the water to have emanated from the sewage-fields of the asylum. For the supposition that the Saale may have washed down the cholera-bacilli from places situated further up is at variance with the circumstance that no cholera-cases occurred above Nietleben. The assumption that the cholera was first brought to the asylum itself in an isolated case, and that this case then became the starting-point from which the infectious matter passed and repassed in a vicious circle through the establishment, therefore seemed the most plausible. This supposed first case, however, was not to be found, and a whole series of hypotheses as to the origin of the Nietleben epidemic, embracing pretty nearly all possibilities and impossibilities from the autochthonous origin of cholera to the itinerant young artisan, who was supposed to have privily deposited his cholera-dejecta on the bank of the Saale, have consequently cropped up. It is not my business here to discuss these hypotheses, and I shall restrict myself to reporting what may contribute most to the clearing up of this dark point.

At the time when the cholera broke out at Nietleben the only cholera-focus from which the disease could be brought was in Hamburg-Altona; but how should it have come from there in the middle of winter, and after skipping so many places that lay between? It could not have been brought by river-vessels, for the river-traffic, which does indeed go from Hamburg up the Elbe and the Saale to above Nietleben, had been in abeyance for about a month and a half owing to the ice. The only other possible bearers one could think of were goods or persons who had come from Hamburg.

As all the traffic of the establishment, including especially the postal traffic, goes through the office, it was soon possible to ascertain that, except some letters, nothing had come to the asylum directly from Hamburg. And, even if goods had really found their way from Hamburg into the asylum, it would not have been admissible to brand them as bearers of cholera, for, as is well known, cholera has never yet been carried by commercial goods proper, and even during the great Hamburg epidemic not a single such case could be proved, though great quantities of goods were sent from Hamburg after the beginning of the epidemic.

It follows that the infection can have been carried only by a person or persons, and this supposition seems to me the most probable. As the asylum receives its patients only from the Prussian province of Saxony, the patients may for the present be left out of account. Unless indeed the cholera had been brought from Hamburg to the province of Saxony by a mild case, such as were observed repeatedly at that time in the beggars' inns of Hamburg, and some person had been infected by this undiscovered case, and sent to the asylum soon after owing to mental disease. Instead of this somewhat complicated way, which, indeed, cannot be designated as impossible, but also not as quite probable, there is another and simpler way by which the disease may have been brought direct from Hamburg, namely, through the staff of attendants, which is always more or less fluctuating. During the three months preceding the epidemic the asylum had engaged 13 attendants of both sexes. None of them gave Hamburg as their last place of abode, but it came out by accident that a male attendant engaged from Halle had come from Hamburg immediately before. He had stayed only a few days in Halle, in order to learn whether he would be employed at Nietleben. In the first days of his service at Nietleben he suffered from severe diarrhœa, which also would not have become known, had he not neglected to report himself in Halle, in obedience to the regulation still in force at the time, as having come from Hamburg. On being required to pay a fine to the police for this omission, he excused himself on the ground that he had felt too unwell to be able to report himself. It cannot be maintained without more ado that just this attendant, who moreover is said to have been in a Hamburg establishment which had remained free of cholera, brought the disease to Nietleben. He suffered from diarrhœa between the 5th and the 8th of December, and there would have been a strikingly long time between the introduction of the first germ of the disease and the outbreak of the epidemic. But this incident at any rate proves that direct personal intercourse took place between Hamburg and Nietleben in the period before the epidemic. The Hamburg epidemic died out at that time; many persons who had gone thither to find employment as attendants on the sick had to look out for other places, and one or other of the eight persons who were engaged for Nietleben in Halle during the period in question may have been in Hamburg shortly before.

I cannot share the conjecture that has been put forward that the cholera was brought to Nietleben in the preceding summer, remained latent there for a length of time, and then suddenly broke out explosively; for I cannot suppose that the infectious matter would have lain inactive somewhere or other, in the ground for instance, during the warm season so favourable to it, and suddenly awakened just at the coldest time, when the ground was frozen to a considerable depth, nor that, under circumstances so extremely favourable to cholera as those of the asylum were, it remained unnoticed for months through a chain of mild cases, in the bodies of the inmates for instance.

After all that has been ascertained regarding the manner

in which cholera is carried from place to place, it still seems to me most probable that it was brought straight from Hamburg by one or more of the attendants.

It is to be regretted that no full certainty could be obtained on this point. It would be a great mistake, however, to conclude from this, as has actually been done, that the real essence of the question which the Nietleben epidemic puts to us has therefore remained unanswered, and that the whole investigation is therefore destitute of importance. It would certainly have been very agreeable, especially for non-experts, if one could have definitely named the person who is supposed to have brought the cholera to Nietleben, but the fact that this person has remained undiscovered does not create a gap involving the whole edifice in ruin. The present epidemic has taught us hundreds of times how cholera is spread by human intercourse, and, if once and away we do not succeed in discovering the connection between various foci of pestilence, we must assume that in such a case also matters have gone just as in the overwhelming majority of cases in which the carrying of the infection was proved. It would be childish to demand that the investigators of cholera should either demonstrate every transmission of cholera under all circumstances and in spite of all the thousand complications of human intercourse, or else abandon their researches altogether. As to the Nietleben epidemic, we may regard it as perfectly certain that the cholera was brought from the Hamburg focus by a person or persons. The important experience gained in this epidemic, and the equally important lessons to be learned from that experience, do not lose one whit of their value because the infecting person himself could no longer be named, just as the Hamburg epidemic is not less instructive to us because the name of the Russian emigrant who carried the first cholera-germ thither cannot be given.

As regards the measures taken at Nietleben for the combating of cholera, I have the following communication to make.

In summer 1892, when cholera broke out in Hamburg, the medical superintendent ordered that only boiled water should be used in all parts of the establishment, and, in order to have sufficient quantities of boiled water always in store, it was arranged that two large copper kettles in the common kitchen, containing 400 litres each, should be kept constantly at work boiling water. At first the boiled water is said to have been fetched from the kitchen in plentiful quantities by the attendants, but in course of time the use of boiled water diminished to such a degree that shortly before the outbreak of the epidemic only two to three kettlefuls, that is 800 to 1200 litres, were fetched in the course of a week.

Immediately after the beginning of the epidemic this measure was again strictly enforced, and one to two kettlefuls of water are said to have been used daily since then.

I cannot regard this arrangement as of any essential use. If people could get all the water they use in a boiled state, then indeed infection by water would be impossible. But, if orders are given to supply boiled water in a lunatic-asylum, while at the same time the patients are permitted access to the water used for bathing, for washing, for rinsing the closets, etc., and it is moreover a question whether all the attendants are conscientious enough to fetch the water from the boiling-place, the choice of the water will be for the most part determined by the better taste of the unboiled water or by the convenience of the attendant, and one must not suppose that such a measure provides a reliable protection against cholera-infection.

Accordingly, when the real cause of the epidemic was discovered, it was resolved to cut off the asylum from the infected conduit-water as soon and as completely as possible. But insurmountable obstacles to the instant execution of this measure arose, and this case showed how extremely difficult it is suddenly to alter a centralised water-supply, which is connected with all the parts of the sphere which it provides, and has become, so to speak, part and parcel of its life and activity. There was no want, indeed, of a substitute for the water to be excluded, for the city of Halle had at once declared its readiness to send twenty to thirty cubic metres of its conduit-water to the asylum daily in carts (which were used under ordinary circumstances for sprinkling the streets, and were very well adapted for the transport of water), and had begun to fulfil its promise on the 20th of January. The quantity of water thus placed at disposal, however, served only to supply the kitchen and the wards, while the steam-boilers, the washing-house, and the closets had to be supplied with the infected water as before; and so there was nothing for it but to close the outlet-cocks in the wards, the kitchen, and other parts of the asylum used for domestic purposes, and leave the ordinary water-supply uninterfered with in all the other parts of the establishment. Very soon, however, it came out that, despite the strictest orders of the medical superintendent, some of the attendants had used the keys to the water-supply still left in their possession, in order to get conduit-water for their wards, evidently to save themselves trouble. It was observed too that asylum-patients went to the closets, and drank the rinsing water there. It was therefore found necessary to shut off the water in the closets, and to take the water-keys from the attendants. It was not till the 25th of January that one could feel assured that the inmates of the asylum really no longer got any of its conduit-water, and from that day forward the epidemic rapidly abated. Of course the shutting off of the conduit-water and the end of the epidemic could not be absolutely simultaneous. The shutting off was, as the above description shows, only gradual, and could moreover have no influence on the cholera-cases which were not caused by the infected water but by direct transmission, that is by secondary infection. But the favourable effect which the shutting off of the conduit-water had is evident from the statistics, and becomes still more so, if we submit the cases that occurred after the 25th of January to somewhat closer inspection.

On the 26th of January three persons fell ill, one of them

in the part of the asylum reserved for uncleanly patients (L I.). It was ascertained that she had drunk conduit-water three days before. The second of these three patients was an asylum-patient who had to carry the soiled linen from the cholera-station to the disinfecting-house, and probably caught the infection in performing this duty.

Two cases occurred on the 28th. One of the patients was again one of the uncleanly asylum-patients of the department designated as L I.; the other was the wife of the chief engineer, resident in the third of the houses for officials. She had been suffering from diarrhœa since the 24th, and had had the water-cock open in her kitchen, and also used the water, till that day.

On the 31st one cholera-case occurred, again in L I.

The next case occurred on the 1st of February, and the patient was a female attendant occupied in nursing the cholera-patients.

The next case, about which nothing particular is noted, occurred on the 4th, and the next on the 5th, again in L I.

Then came a longish interval, after which, on the 10th of February, a quite isolated case occurred, between which and the previous cases no connection whatever could at first be traced, till it was discovered to the general surprise that, despite all the precautions that had been taken, the patient had got access to the conduit-water. An attendant had turned on the rinsing water of the urinal with the key of the gas-conduit, which happened to fit the water-conduit too, and a stoker who was accidentally there at the time had seen the patients catching the water with their hands and drinking it.

In consequence of this occurrence the medical superintendent had all the water-outlets soldered up.

The last case occurred on the 13th of February, and the patient was a female attendant in the cholera-station. This case, like those of the other female attendants of that station, was probably due to direct infection.

Though there were doubts at the beginning of the epidemic whether the disease really was Asiatic cholera, the medical superintendent, duly recognising the danger, had taken all measures to prevent the spread of the pestilence to the neighbourhood so early as the second day. No patients were dismissed or received. The chief administrative official of the district ordered that attendants who wished to leave the establishment should name their future place of residence. In such a case the competent local authority was to be informed of the arrival of the persons suspected of cholera, in order that they might be subjected to medical observation for five days. Unnecessary visits were not permitted during the epidemic, and outsiders who had to enter the asylum were forbidden to eat or drink there. Other limitations of intercourse than these slight and perfectly reasonable ones were certainly not necessary. But, contrary to my advice, some further and quite harmless limitations were imposed by the city of Halle, evidently with a view to reassuring the inhabitants, who had been alarmed by the calamity which had so suddenly befallen Nietleben.

The judicious quartering and isolating of the numerous cholera-patients was fortunately a matter of no insuperable difficulty. Two new pavilions (U and W); situated apart from the other buildings, happened to be still unoccupied, and the circumstances admitted of the evacuation of a third pavilion (V) belonging to the same group of buildings, so that three houses were at disposal. The middle one (W) was arranged as a cholera-hospital, one side for men, the other for women. The other two served as observation-stations for persons suspected of cholera (U for women, V for men).

The detection of the patients affected with suspicious symptoms was not quite easy at first, for the asylum-patients refrained for the most part from reporting themselves ill, and the attendants noticed only the more striking cases of disturbed digestion. All the patients, including those who seemed quite well, were therefore required to use night-stools only, a regulation which had proved very useful in Hamburg under similar circumstances. Thanks to this regulation, even the slightest cases of diarrhœa no longer escaped observation. These cases of course could not but be regarded as suspicious, and, as they were numerous, the observation-stations soon became alarmingly full. In Hamburg the suspected were always bacteriologically examined at once, and those who were not suffering from cholera dismissed as soon as possible, whereby the overcrowding of the observation-stations was easily avoided. At Nietleben, on the other hand, bacteriological investigation was impracticable at first for want of apparatus and assistants. It was not till the 1st of February that a bacteriological laboratory, established for this purpose in the asylum with all possible speed, was ready, and all suspicious dejecta could be regularly investigated. All the patients in the observation-station whose dejecta were free of cholera-bacteria were lodged first for some days at an intermediary station, and sent back to their own wards only if they remained well there too. The bacteriological investigation very soon emptied the observation-stations at Nietleben also, and at the same time afforded that sure knowledge of the real state of the epidemic which was so desirable. Before the 1st of February the great number of the suspected occasioned no small disquietude, but from that day forward it was possible to survey the course of the epidemic,—a matter of special importance when an epidemic is approaching its close.

At Nietleben, just as in the Hamburg after-epidemic, the bacteriological investigation detected several cholera-cases which could not indeed be reckoned among the slightest, but yet among those which, owing to the slightness of the diarrhœa, would not have been recognised with certainty as cholera by their clinical symptoms, and could at most have been designated as slight cholerine. It is also noteworthy that cholera-bacteria were found in the evacuations of two convalescents so late as about three weeks after they fell ill. Unfortunately these patients were not bacteriologically examined from the first, so that the period during which the cholera-bacteria were able to maintain themselves in their intestines could not be ascertained to a day. But they confirm an observation of high prophylactic importance which had been made elsewhere, and teach that the danger of infection does not always cease simultaneously with the cholera-attack proper.

Besides the isolation of the cholera-patients, the disinfection of the evacuations and of all objects soiled with them was as far as possible effected. The linen and clothes were disinfected in steam-apparatuses, the fluid excrements partly by means of solutions of carbolic-acid-soap, partly by means of quick-lime.

Towards the end of the epidemic, when only isolated cases occurred, the disinfecting of the sick-rooms could also be undertaken. For this purpose the several rooms or whole wards were evacuated, the walls, floors, bedsteads, etc. washed with a disinfecting fluid, the rooms copiously ventilated, and then re-occupied by the patients, who had meanwhile been bathed and provided with disinfected linen, clothes, and bedding. If nevertheless a new case occurred, as, for instance, the above described cases did on the 5th and 10th of February, the whole procedure was repeated.

The water-supply and the sewage-fields also demanded special attention. As it is not yet known how long the cholera-bacteria can remain viable in water or in the soil even in winter, the water-supply could not be restored to use without being disinfected. It seemed necessary to disinfect the sewage-fields also, for cholera-bacteria had been found in them repeatedly and at different places, and it was to be feared that, when thaw set in, they would be washed in considerable quantities with the melted water into the Saale.

The disinfecting of the water-conduit was not a matter of insuperable difficulty. Diluted lime-milk, a solution of carbolic acid, or a mineral acid might have been used for that purpose. Carbolic acid was chosen, and a three-percent. solution of it was driven from the pump-shaft into all parts of the conduit, left there for 24 hours, and then washed out with Halle water. We are justified, I think, in assuming that a reliable disinfection was thus effected.

The fact that the conduit-water had an unpleasant taste of carbolic acid for a good while after might be adduced as a disadvantage of this method; but carbolic acid is preferable to lime-milk, because it enables one to avoid with certainty the choking-up of the pipes, and to mineral acids, because it cannot injure their inner surfaces.

The disinfecting of the sewage-fields was a much harder task. The disinfecting of the whole surface was out of the question, owing to the very large quantity of the disinfectants that would have been needed. Nor was it necessary, for the sewage had not spread over the whole field, but had only flowed in definite narrow channels. Lime-milk was therefore added to the sewage in large quantities till the reaction of the fluid in the main drain at the lower end of the field was strongly alkaline. Some time afterwards, when the sewage-fields were researched for cholera-bacteria, none were found. They had disappeared; whether in consequence of the disinfection, or of climatic influences—which seems to me more likely,—may remain an open question.

Though all the excrements of the patients were disinfected as promptly as possible, undisinfected dejecta must, quite at the beginning of the epidemic, have flowed into the sewage and with it into the Saale. Moreover it was doubtful whether all the infectious matter was reliably annihilated by the disinfection, and whether it was impossible that cholera-bacteria got into the Saale even later. If these doubts were well founded, it was not impossible that the infectious matter had been carried on by the Saale, and that outbreaks of cholera would develop in the places situated further down that stream. It was necessary to take steps as promptly as possible to prevent the spreading of the pestilence in this way. The almost simultaneous occurrence of cholera-cases at several places on the Saale, and the discovery of cholera-bacteria in the sewage-fields and in the water of the Saale below Nietleben, very soon showed how well founded these apprehensions were.

But how was the propagation of cholera by the water of the Saale to be prevented?

One cannot disinfect rivers; at best one can draw the attention of the people who live on their banks to the dangers connected with the use of the river-water. This was done as impressively as possible; the competent authorities not only warned the people against using the water of the Saale, but positively forbade it. They were well aware, indeed, that the prohibition could not really be enforced, and that its only real effect would be to make the danger appear to the population as one of special gravity. The greatest importance was attached not to the avoidance of infection, which could not be avoided, but to doing all that could be done to render new cholera-foci harmless as quickly as possible. For this purpose medical men, clergymen, teachers, heads of communities, gendarmes, in short all persons who might be expected to take an interest and especially an intelligent interest in the situation, were called upon by an order of the district-government at Merseburg, valid for all places situated on the Saale, to notify every case of illness that looked at all like cholera.

Very soon after the publication of the order such notifications were received, and every case was at once examined on the spot by the medical officer of the Saale district, and bacteriologically investigated in the Hygienic Institute at Halle, in order that the real cholera-cases might be detected as quickly as possible. We have good reason to believe that this purpose was everywhere attained, and that the authorities were thus enabled energetically to attack the cholera-foci in the initial stage of their development by all possible means.

The first place at which cholera appeared was Trotha, a village on the right bank of the Saale, five kilometres (a little more than three miles) below Halle. Three persons fell ill of cholera there on the 24th of January, and one of them died. A fourth case, due to secondary infection, ensued on the 29th.

On the 28th a cholera-case was detected at Wettin on the right bank of the Saale, twenty kilometres below Halle. The patient was a woman, who died some days later. She seems to have fallen ill on the 24th, but had not sent for a doctor till the 28th, when her condition became worse. Researches set on foot at Wettin did not lead to the discovery of any other indubitable cases of cholera. In one case of illness, which looked exceedingly suspicious, evacuations for bacteriological investigation were no longer to be had.

At the village of Cröllwitz on the left bank of the Saale, two to three kilometres below Nietleben, a suspicious case was notified on the 30th of January, and ascertained to be cholera on the same day. This was the first of six cases, two of which proved fatal. As five of the patients were members of one family, who did not fall ill simultaneously but at intervals of several days, the infection in some of these cases was doubtless secondary.

The last of these outbreaks happened at another village on the left bank of the Saale, Lettin (six to seven kilometres below Nietleben), where three cases were ascertained on the 2nd, 3rd, and 4th of February, one of which proved fatal.

If one seeks out these places on the map (diagram 2), where they are underlined, it strikes one at once that they are all situated below Nietleben, some on the left, some on the right bank of the Saale. They are not connected with one another by roads or other lines of communication, and it has been proved that they have not the slightest mutual intercourse. The only possible medium of intercourse, the barge-traffic on the Saale, was completely excluded, for, except at some places where the current is unusually strong, the river was thickly covered with ice. None of the places had any connection with the Nietleben Asylum. The only thing they have in common is the use of the Saale water, and it was easily proved that, despite the above-mentioned strict prohibition, the water really was used in all of them.

At Wettin holes had been made in the ice before the houses situated on the Saale, to make the water accessible; and the carrying of water from the Saale into the houses could be prevented at last only by setting policemen to watch the bank.

At Cröllwitz the accessible parts of the river had to be enclosed with board-fences for the same reason. Moreover the patients at both places frankly confessed that they had used Saale water for domestic purposes.

At Trotha and Lettin, on the other hand, it was maintained that the prohibition had been rigorously enforced, and that nobody had taken water from the Saale. And yet at both places the Saale water had played a part in the matter. The water had not, indeed, been fetched directly from the river, but it was proved that the patients had drunk from water-pipes conveying Saale water for the watering of cattle. At Lettin there was such a conduit for the sheep-farm of the domain, and the shepherd and a farm-servant, both of whom were attacked by cholera, had both drunk from it. In the dwelling of the shepherd, whose illness proved rapidly fatal, a child of his evidently caught the infection directly.

The cholera-outbreak at Trotha was of a peculiarly interesting nature. The disease entirely restricted itself there to one house. But this house was a kind of workmen's barrack, inhabited by fourteen families consisting of 62 persons. Each family had one or at most two rooms at its disposal. The breaking out of cholera among this densely crowded mass of human beings, living under the most unfavourable hygienic conditions, was the more alarming as the same house had been most severely attacked by cholera in a previous epidemic in 1866, and sixteen to eighteen deaths from cholera (I could not ascertain the exact number) are said to have occurred in it on that occasion. At the very first investigation the striking circumstance was noted that three of the numerous inmates had fallen ill in quite different parts of the house, one on the ground floor, another in the middle story, the third in the upper story, and that the patients were all men; women and children were entirely spared at first. The further inquiry yielded the following result. These workmen's families, most of them from Upper Silesia, lived quite by themselves, had no intercourse with the city of Halle or the Nietleben Asylum, and but little even with the other inhabitants of the village. The patients had been employed in feeding the fattened cattle at a sugar-factory to which the house in question belonged, and here again there was a conduit conveying Saale water to the cattle-stalls. A branch-pipe of this conduit, indeed, went into the house, and, under ordinary circumstances, all its inmates used the Saale water. Just at the time in question, however, the branch-pipe was frozen, and the Saale water went to the cattle-stalls only. The women and children, therefore, were forced to use the water of the neighbouring wells, whereas the men, according to their own account, had drunk the conduit-water in the stalls. The one especially who suffered most severely, and died of cholera, is said to have drunk water in copious draughts after eating a large quantity of horse-sausage.

What I have here reported of the cholera-outbreaks in the places on the Saale, and especially the behaviour of the disease in the individual outbreaks, at Trotha for instance, admits of only one interpretation, namely, that the infectious matter of cholera, whatever one may suppose it to be, was carried from place to place by water. In what other manner could the pestilence have found its way in hard midwinter, immediately after the Nietleben outbreak, into sequestered places, connected in no other way either with each other or with the cholera-focus, restricting itself moreover exclusively to places situated on the river? Whoever, in the face of these facts, persists in denying that water may be the bearer of the infectious matter of cholera is not accessible to the logic of facts at all.

From the circumstance that the cholera did not go beyond Wettin, which lies thirteen to fourteen miles below the starting-focus, it might perhaps be inferred that, in winter at least, the cholera-bacteria cannot be washed to greater distances in a viable condition ; but this question cannot be decided by the instance before us for the following reason. At Friedeburg, a few kilometres below Wettin, the Schlenze, a little river coming from the west, flows into the Saale, bringing with it the water from the Mansfeld Mines with the Key-Gallery. The water of those mines, owing probably to the lixiviation of subterranean salt-beds, has for some time past been so rich in salt that the water of the Schlenze contains about ten per cent. of common salt at the place where it flows into the Saale. In consequence of this the Saale water below Friedeburg is so salt that it is of no use, at least for domestic purposes. There it is not necessary to station gendarmes on the bank, or to erect fences, in order to keep away the people from the river. From there downwards the water prohibits its use itself, and it is very probable that the non-appearance of cholera at the places below Friedeburg was due to this circumstance much more than to the dying of the cholera-bacteria.

At all the places on the Saale where cholera broke out the measures which had proved so efficacious the year before were energetically enforced. The patients were isolated ; the members of their families and other persons open to the suspicion of being infected were carefully observed, and bacteriologically examined on the appearance of the slightest disturbances of digestion, and the excrements, linen, and sick-rooms of the patients were disinfected.

Special difficulties had to be contended with at Trotha, where shelter had suddenly to be found for about fifty people, whom it was impossible to leave any longer packed together in the workmen's barracks. The only way out of the difficulty was to arrange the school-house for this purpose, and this was done with all possible promptitude. Only the patients and their families remained in the house. Among the latter only one fresh case occurred, that of a boy who had evidently caught the infection at the very first in helping to nurse his father, who was severely ill of cholera, in a dirty little room. Among the inmates who were taken to the school-house not a single case occurred.

Thanks to the indefatigable exertions that were made on all hands, among which the self-sacrificing labours of the chief administrative official and the medical officer of the Saale sub-district deserve quite special recognition, the new cholera-foci were all nipped in the bud ; and the spread of cholera, first down the Saale from Nietleben, and then, as was to be expected if these efforts had not been made, over wider districts of the province of Saxony, was prevented.

If one has to occupy oneself with the Nietleben cholera-epidemic, the question forces itself upon one whether this calamity, which cost a not inconsiderable number of human beings their lives, could not have been prevented. Certainly it could. All that was necessary was to take care that the in themselves expedient sanitary institutions of the establishment, the waterwork with its filters and the sewerage with the sewage-fields, performed their functions properly. I should like to protest beforehand against any attempts that may possibly be made to derive from the failure of the Nietleben filters and sewage-fields arguments against the expediency of these institutions in general. In my essay on "Water-Filtration and Cholera" I have already said what I have to say as to how one must judge in future of water-filtration in connection with our latest experience of cholera; and, with reference to the irrigation-method, I wish to say that I regard the treatment of sewage by means of irrigation, judiciously carried out, as the best cleansing method we yet possess. The failure of the filters and the sewage-fields to do what was expected of them was due not to themselves but to their defective treatment. It seems to follow that those under whose management and superintendence the sanitary institutions of the establishment were placed must be made responsible. But I do not believe that this would be just. Not individual persons are to be blamed for the disaster in question, but the circumstances under which we are now living.

One cannot possibly require that the medical superintendent of a lunatic-asylum or the technical official of the government should, in addition to the special knowledge demanded of them, be better hygienists than are some professors of hygiene, whose insight into the finer processes of filtration in sand-filters and in the soil is still defective.

In general the demands on the hygienic responsibility of the medical superintendents of asylums and the like must not be exaggerated. There are certain kinds of knowledge which are not acquired in the ordinary course of hygienic study calculated for medical practitioners, and which also cannot be got out of books, but only by special studies and by experience gained in practice. In this province of things the responsibility of medical men who have undergone the ordinary hygienic training ceases, and one has just as little right to call the medical superintendent of an asylum to account for an explosion of cholera due to mistakes that have been made in filtration and irrigation as for the explosion of a steam-boiler in his establishment in consequence of some easily recognisable and avoidable mistake.

There is only one remedy, a remedy to which I already pointed on a former occasion, and which I should like to advocate once more as urgently as possible here, namely, the State-Supervision of such works by special experts, thoroughly versed in all the circumstances connected with them, and possessed of the necessary experience which constant practical contact with them affords.

But will the State consent to undertake this task? Judging from what I can see of the circumstances, I do not believe that it will do so soon. The resolution will certainly have to be taken sooner or later, but the whole question is not yet considered ripe for discussion. Among persons in whose hands the decision lies one always hears the opinion that men of science are not yet agreed among themselves, and that it is therefore impossible as yet to assume a definite attitude towards this question. The bacteriologists, they say, indeed maintain that cholera and enteric fever can be spread by water, but others whose authority is not inferior to theirs dispute that, and it is not even known yet whether the cholera-bacteria really are the cause of cholera, and whether, in the combating of cholera, they deserve to have such attention paid them as the bacteriologists advise. How deeply rooted such opinions are is most clearly evident from the fact that the principle was laid down not long ago that the chairs of hygiene should be filled alternately by hygienists who are at the same time bacteriologists and by hygienists of the opposite way of thinking, that is, of course, hygienists who attach no importance to bacteriology.

But who are the men of science who are said not to agree as to the importance of the cholera-bacteria? Of course they can only be people who have occupied themselves with bacteriology, that is the so-called bacteriologists. Now I can maintain with certainty that there is no bacteriologist of name existing who does not admit the cholera-bacteria to be the immediate cause of cholera. Even the Munich School, which persisted longest in opposition, has had quite gradually to consent to allow them at least the part of X in the well-known equation with three unknown quantities. The only question as to which the men of science who are alone competent to judge of this matter differ is, what further auxiliary factors, acting in the human body and outside of it, are to be admitted, and to what extent. But, as regards the real main question, men of science are completely at one.

Those men of science, then, who pooh-pooh the cholera-bacteria are not bacteriologists; their science has its roots in some other field. But they have a great advantage in the discussion of the cholera-question. They do just as other people do who understand nothing of a matter; they speak about it with a definiteness and certainty which cannot but powerfully impress and has always hitherto powerfully impressed the non-expert, that is in this case the non-bacteriologist. By the medical public and by the official bodies which have to do with cholera they are therefore regarded as authorities, as "men of science," who have not yet arrived at agreement with the other men of science.

There are no tokens yet to justify the hope that non-bacteriologists will cease to put in their word on these questions, and to bring ever new confusion into the minds of the public at large. Von Pettenkofer at least, who, as he himself emphasizes on every occasion, has not occupied himself with bacteriology, declared in his very last publication against the standpoint now assumed by all bacteriologists, and even by his own pupils,* and dismissed the bacteriological side of the cholera-question with jokes about "bacillus-catching" and the "impossibility of making human intercourse fungus-proof," though he ought to know very well that the measures now taken against cholera are not based on the principle of making human intercourse fungus-proof. It is to be hoped that, after our experience in the last epidemic of the measures so obstinately combated by him,† he has already convinced himself that they are not so bad after all as he imagined.

Should Von Pettenkofer nevertheless persist in his attitude of opposition, I should understand that, not indeed from the scientific but from the human point of view. It must be extremely difficult for him, bound up and grown old with the opinions advocated by him during a long series of years with the greatest expenditure of genius and sagacity, to sever himself from them, at least in part. But it is unintelligible to me that a man like Liebreich, who also has not occupied himself with bacteriology, and who in fact, as almost every sentence of his recent lecture to the Berlin Medical Society,‡ proves, does not understand bacteriology at all, who, moreover, has evidently never made even a single bacteriological investigation of cholera-dejecta himself, can undertake to pronounce a verdict of condemnation against the bacteriological cholera-diagnosis in particular, and against bacteriology and its past achievements in general. What is likely to be the result if, on the one hand, the men of science in the

* Emmerich, who made the well-known experiment of swallowing cholera-bacteria along with Von Pettenkofer, has in a recent publication admitted the cholera-bacteria to be the cause of cholera.

† I mean of course only those measures which were advised by medical men, not those which were ordered by some official bodies in their overzeal over and above. As regards the latter I entirely agree with Von Pettenkofer.

‡ German Medical Weekly, 1893, No. 26.

department of bacteriology take all imaginable pains to prove that the purity of filtered water must be bacteriologically tested, while, on the other hand, the man of science Liebreich declares: "Bacteriology has thrown no new light on the water-question; good water was demanded in former times as well as now; we knew long ago that putrid water makes people ill." Is that not sowing confusion with all one's might?

I fear that, so long as such words are spoken, people in power will keep always repeating: "Men of science are not agreed yet, and everything must remain as it is for the present." But if then, as I also fear, such catastrophes as those of Hamburg and Nietleben continue to befall us, people will have to thank those "men of science" for it, who vindicate for themselves the extremely responsible office of talking of things they know nothing about.

www.ingramcontent.com/pod-product-compliance
Lightning Source LLC
Chambersburg PA
CBHW030310170426
43202CB00009B/952